V International Conference la ValSe-Food and VIII Symposium Chia-Link

V International Conference la ValSe-Food and VIII Symposium Chia-Link

Editors

Claudia Monika Haros
Loreto A. Muñoz
María Dolores Ortolá

Basel • Beijing • Wuhan • Barcelona • Belgrade • Novi Sad • Cluj • Manchester

Editors
Claudia Monika Haros
Spanish Council for Scientific
Research (IATA-CSIC)
Valencia
Spain

Loreto A. Muñoz
Universidad Central de Chile
Santiago
Chile

María Dolores Ortolá
Universitat Politècnica de
València
Valencia
Spain

Editorial Office
MDPI
St. Alban-Anlage 66
4052 Basel, Switzerland

This is a reprint of articles from the Proceedings published online in the open access journal *Biology and Life Sciences Forum* (ISSN 2673-9976) (available at: https://www.mdpi.com/2673-9976/25/1).

For citation purposes, cite each article independently as indicated on the article page online and as indicated below:

Lastname, A.A.; Lastname, B.B. Article Title. *Journal Name* **Year**, *Volume Number*, Page Range.

ISBN 978-3-03928-625-6 (Hbk)
ISBN 978-3-03928-626-3 (PDF)
doi.org/10.3390/books978-3-03928-626-3

Cover image courtesy of Claudia Monika Haros

Contents

About the Editors

Claudia Monika Haros

Claudia Monika Haros holds a PhD in Chemical Sciences from the University of Buenos Aires, Argentina. She has developed her professional work in Universities and National and Foreign Research Centers since 1990. Currently, she is the Head of the Cereals Group of the Institute of Agrochemistry and Food Technology belonging to the Higher Council for Scientific Research of Spain (IATA- CSIC). She is a founding member of the Chia-Link Network, which is included in the ValSe-Food Network (Iberoamerican Valuable Seeds or Valiosas Semillas Iberoamericanas) that she coordinates. Her research on ancestral Latin American crops aims to work in an environment of cooperation with the international scientific community and unite efforts to promote safe, sustainable, tasty, nutritious and healthy foods with cereals/pseudocereals or their by-products, among other crops such as chia; this is through collaborations between the sectors involved in research, academic institutions, industry and society.

Loreto A. Muñoz

Loreto A. Muñoz holds a PhD in Food Science and Engineering from the University of Santiago de Compostela and a PhD in Engineering Sciences from the Pontificia Universidad Católica de Chile, a Master's degree in Food Science from the University of Chile and is a Food Engineer at the University of Santiago de Chile. In 2012, she received the BIMBO Pan-American Award for her research on chia seeds. She is currently a research academic and Director of the LabCial Food Science Laboratory of the School of Engineering at the Central University of Chile and International Coordinator of the CHIA-LINK Network. Her research has been focused on the extraction and characterization of food materials of plant origin such as seeds, grains and legumes in terms of their microstructural, physicochemical and functional properties, as well as the study of digestibility, bioaccessibility and associated health effects.

Claudia Monika Haros

María Dolores Ortolá is a full professor at the Universitat Politècnica de València (Spain) in Food Technology, a professor at the Department of Food Technology of the UPV and researcher at the Institute of Food Engineering-FoodUPV. Their teaching and research activity is focused on the design of equipment and facilities for agri-food industries, food reformulation with better nutritional properties, and post-harvest technologies for fruits and vegetables. In these three fields, they have participated in competitive research projects, published articles in indexed journals, and supervised 13 doctoral theses. One of the lines of research on which they have worked in recent years is the study of the adaptation of new crops such as Moringa Oleifera to the Mediterranean climate and its revaluation (especially its leaves, pods, and seeds) as a food ingredient, especially for its high content of proteins and essential amino acids.

Preface

Iberoamerican crops are currently underutilized, with low cultivation levels. However, there has been a notable increase in the global demand for these crops, leading to a surge in their production and prices. Over the years, these valuable seeds have garnered recognition from food scientists and producers due to their exceptional nutritional value. They boast high-quality proteins, significant amounts of starch and/or fiber with unique characteristics, and abundant micronutrients such as minerals, vitamins, and bioactive compounds. Additionally, their gluten-free nature makes them suitable for individuals with gluten intolerances or allergies.

The growing interest in Iberoamerican Valuable Seeds has led to intensified research efforts in recent decades. This book encapsulates the proceedings of the V International Ia ValSe-Food and VIII Symposium of Chia-Link Network, entitled "Feeding Biodiversity and Mitigating the Effects of Climate Change: the Role of Ancestral Crops in Creating Healthy Food," which took place at the University-Business Foundation of the University of Valencia (ADEIT) from October 4 to 6 in Valencia, Spain.

The structure of the conference was divided into different sections to cover various topics related to agriculture, food technology, nutrition, health promotion, and climate change.

Session I: Agronomy and Crop Diversity

Session II: Climate Change and Food Systems; Innovations in Food Science and Technology; Sustainable Management of Food Waste

Session III: Research in Food and Nutrition; Food Chemistry and Bioactive Components of Foods; Food Immunology

Each section included keynote presentations, research paper presentations, panel discussions, and interactive sessions to facilitate knowledge exchange and collaboration among participants. The conference addressed a wide range of topics, fostered interdisciplinary discussions, and promoted holistic approaches to sustainable agriculture, food technology, nutrition, health promotion, and climate change.

Within these pages, readers will find a compilation of recent investigations conducted by the ValSe-Food Group on valuable seeds and other crops, such as the following: amaranth (*Amaranthus* spp.), quinoa (*Chenopodium quinoa*), kañiwa (*Chenopodium pallidicaule*), chia (*Salvia hispanica* L.), Andean maize (*Zea mays* L.), tarwi (*Lupinus mutabilis*), broad beans (*Vicia faba* L.); Mistol (*Sarcomphalus mistol*), bean (*Phaseolus vulgaris* L.), peanut (*Arachis hypogaea*), moringa (*Moringa oleifera*), and kurugua (*Sicana odorifera*). The book offers comprehensive and up-to-date knowledge across various fields of food science, covering production, utilization, structure, and chemical composition. Special attention is given to the carbohydrates, fibers, bioactive compounds, proteins, and lipids found in kernels and other plant parts. The content delves into the processes involved, various food products and applications, as well as the nutritional and health implications associated with these crops.

Claudia Monika Haros, Loreto A. Muñoz, and María Dolores Ortolá
Editors

biology and life sciences forum

MDPI

Editorial

Statement of Peer Review

Claudia M. Haros [1,*], Loreto A. Muñoz [2] and María Dolores Ortolá [3]

[1] Cereal Group, Instituto de Agroquímica y Tecnología de Alimentos, Spanish Council for Scientific Research (IATA-CSIC), 46980 Valencia, Spain

[2] Instituto de Investigación y Postgrado, Facultad de Medicina y Ciencias de la Salud, Universidad Central de Chile, Santiago 8370292, Chile; loreto.munoz@ucentral.cl

[3] Instituto Universitario de Ingeniería de Alimentos-FoodUPV, Universitat Politècnica de València, 46022 Valencia, Spain; mdortola@tal.upv.es

* Correspondence: cmharos@iata.csic.es; Tel.: +34-963900022

Citation: Haros, C.M.; Muñoz, L.A.; Ortolá, M.D. Statement of Peer Review. *Biol. Life Sci. Forum* **2023**, *25*, 14. https://doi.org/10.3390/blsf2023025014

Published: 28 September 2023

In submitting conference proceedings to the *Biology and Life Sciences Forum*, the volume editors of the proceedings certify to the publisher that all papers published in this volume have been subjected to peer review administered by the volume editors. Reviews were conducted by expert referees according to the professional and scientific standards expected of a proceedings journal.

- Type of peer review: single-blind; double-blind; triple-blind; open; other (please describe): single-blind.
- Conference submission management system: The conference was organized through the web: https://congreso.adeituv.es/valse23/?lang=en.
- Number of submissions sent for review: all of them (17).
- Number of submissions accepted: all of them (17).
- Acceptance rate (number of submissions accepted/number of submissions received): all accepted (17/17).
- Average number of reviews per paper: two or three.
- Total number of reviewers involved: three.
- Any additional information on the review process:

Reviewers' Criteria: The editors were the reviewers who were provided with guidelines and criteria by the conference organizers. These criteria include assessing the significance of the research, the rigor of the methodology, the clarity of the presentation, the originality of the work, and its relevance to the field. Reviewers also comment on the strengths and weaknesses of the manuscript and may suggest improvements or revisions.

Editorial Decision: Based on the feedback from peer reviewers, the editors make a decision. The possible decisions include:

- Acceptance: The manuscript is accepted for publication without major revisions.
- Minor Revisions: The manuscript needs minor changes or clarifications before acceptance.
- Major Revisions: The manuscript requires substantial revisions and needs to be reevaluated after revisions.
- Rejection: The manuscript does not meet the necessary quality and is not suitable for publication in its current form.

Revisions and Resubmission: If revisions are requested, the authors make the necessary changes to address the editors' comments and concerns. The revised manuscript is then resubmitted to the editors for further evaluation.

Final Decision: The editorial or handling editor makes a final decision based on the revised manuscript and feedback from the authors. This decision may lead to acceptance, further revisions, or rejection.

biology and life sciences forum

Editorial

Preface of the V International Conference la ValSe-Food and VIII Symposium Chia-Link

Claudia M. Haros [1,*], Loreto A. Muñoz [2] and María Dolores Ortolá [3]

[1] Cereal Group, Instituto de Agroquímica y Tecnología de Alimentos, Spanish Council for Scientific Research (IATA-CSIC), 46980 Valencia, Spain

[2] Instituto de Investigación y Postgrado, Facultad de Medicina y Ciencias de la Salud, Universidad Central de Chile, Santiago 8370292, Chile; loreto.munoz@ucentral.cl

[3] Instituto Universitario de Ingeniería de Alimentos-FoodUPV, Universitat Politècnica de València, 46022 Valencia, Spain; mdortola@tal.upv.es

* Correspondence: cmharos@iata.csic.es; Tel.: +34-963900022

Citation: Haros, C.M.; Muñoz, L.A.; Ortolá, M.D. Preface of the V International Conference la ValSe-Food and VIII Symposium Chia-Link. *Biol. Life Sci. Forum* **2023**, *25*, 15. https://doi.org/10.3390/blsf2023025015

Published: 28 September 2023

1. Introduction

The V International Conference la ValSe-Food and VIII Symposium Chia-Link: Feeding Biodiversity and Mitigating the Effects of Climate Change: the Role of Ancestral Crops in Creating Healthy Food is organized by the International la ValSe-Food (Iberoamerican Valuable Seeds or *Valiosas Semillas Iberoamericanas*) Network—CYTED. The conference was held from 4 to 6 October 2023 in Valencia, Spain.

Ancient grains refer to a category of grains that have been cultivated for thousands of years and have remained largely unchanged by modern plant breeding practices. These grains have a long history as staple foods in various cultures around the world. Ancient grains have gained popularity in recent years due to their nutritional benefits, unique flavours, and potential health advantages compared to modern grains. They are often sought after for their higher fibre, vitamins, and mineral content, as well as their potential to be more easily digested by sensitive populations.

In addition to promoting safe, sustainable, nutritious, and healthy food, la ValSe-Food Network also recognizes the importance of biodiversity and addressing climate change. The network acknowledges that ancient crops play a crucial role in preserving biodiversity and enhancing resilience in agricultural systems.

By studying and promoting ancient seeds, the network aims to contribute to the conservation of genetic diversity. Ancient grains possess unique, naturally selected genetic traits over centuries, making them valuable resources for future breeding programs and the development of climate-resilient crops. Preserving and utilizing ancient seeds can help to maintain biodiversity and protect against the loss of valuable genetic resources in the face of climate change and other environmental challenges.

The aim of la ValSe-Food Network is to establish a framework which brings together scientific, technical and industrial groups who work in all areas related to ancient crops (Figure 1).

The overall objective of the Network is to create a cooperative environment with the international scientific community to promote safe, sustainable, tasty, nutritious and healthy food with ancient crops via collaborations among the sectors involved in it: research, institutions, industry and society.

In addition, the aim of the International la ValSe-Food Network is to promote the sustainable development of science and technology based on the study of ancient seeds.

Figure 1. Scientific, technical and industrial groups involved in la ValSe-Food Network.

2. Topics

The conference was structured into different sections to cover various topics related to agriculture, food technology, nutrition, health promotion, and climate change.

- **Session I:** Agronomy and Crop Diversity
- **Session II:** Climate Change and Food Systems; Innovations in Food Science and Technology; Sustainable Management of Food Waste
- **Session III:** Research in Food and Nutrition; Food Chemistry and Bioactive Components of Foods; Food Immunology

Each section included keynote presentations, research paper presentations, panel discussions, and interactive sessions to facilitate knowledge exchange and collaboration among participants. The conference addressed a wide range of topics, fostered interdisciplinary discussions, and promoted holistic approaches to sustainable agriculture, food technology, nutrition, health promotion, and climate change.

3. Committees

3.1. Conference Chair

Claudia Mónika Haros
Coordinator of la ValSe-Food
Cereal Group, Instituto de Agroquímica y Tecnología de Alimentos, Spanish Council for Scientific Research (IATA-CSIC), Valencia, Spain

3.2. Conference Co-Chairs

Loreto A. Muñoz
Coordinator of Chia-Link Network
Instituto de Investigación y Postgrado, Facultad de Medicina y Ciencias de la Salud, Universidad Central de Chile, Santiago, Chile

Mª Dolores Ortolá Ortolá
Instituto Universitario de Ingeniería de Alimentos-FoodUPV, Universitat Politècnica de València, Valencia, Spain

3.3. Scientific Committee

Claudia Mónika Haros
Coordinator of la ValSe-Food
Cereal Group, Instituto de Agroquímica y Tecnología de Alimentos, Spanish Council for Scientific Research (IATA-CSIC), Valencia, Spain

Francisco Millán
Instituto de la Grasa, Spanish Council for Scientific Research (IG-CSIC), Seville, Spain

Javier Fontecha
Instituto de Investigación en Ciencia de la Alimentación, Spanish Council for Scientific Research (CIAL-CSIC), Madrid, Spain

Jose Moises Laparra
Instituto Madrileño de Estudios Avanzados (IMDEA-Food), Madrid, Spain

Loreto A. Muñoz
Coordinator of Chia-Link Network
Instituto de Investigación y Postgrado, Facultad de Medicina y Ciencias de la Salud, Universidad Central de Chile, Santiago, Chile

Ma Carmen Millán Linares
Instituto de la Grasa, Spanish Council for Scientific Research (IG-CSIC), Seville, Spain

Mª Dolores Ortolá Ortolá
Instituto Universitario de Ingeniería de Alimentos-FoodUPV, Universitat Politècnica de València, Valencia, Spain

Maria Reguera
Autonomous University of Madrid, Madrid, Spain

Octavio Paredes-López
Centro de Investigación y de Estudios Avanzados (CINVESTAV), Irapuato, Mexico

3.4. Organizer Committee

Ana Ribera Castelló
Cereal Group, Instituto de Agroquímica y Tecnología de Alimentos, Spanish Council for Scientific Research (IATA-CSIC), Valencia, Spain

Andrea Alonso Álvarez
Cereal Group, Instituto de Agroquímica y Tecnología de Alimentos, Spanish Council for Scientific Research (IATA-CSIC), Valencia, Spain

Claudia Mónika Haros
Cereal Group, Instituto de Agroquímica y Tecnología de Alimentos, Spanish Council for Scientific Research (IATA-CSIC), Valencia, Spain

Mª Dolores Ortolá Ortolá
Instituto Universitario de Ingeniería de Alimentos-FoodUPV, Universitat Politècnica de València, Valencia, Spain

Mª Luisa Castelló
Instituto Universitario de Ingeniería de Alimentos-FoodUPV, Universitat Politècnica de València, Valencia, Spain

Sofía Blanco Haros
University of Valencia, Valencia, Spain

Funding: This Conference was financed by the Project Ia Valse Food-CYTED (Ref. 119RT0567) and CIAORG/2022/123 from a Conselleria d'Innovació, Universitats, Ciència i Societat Digital (Valencia, Spain). The Accreditation of IATA-CSIC as Center of Excellence Severo Ochoa CEX2021-001189-S financed by MCIN/AEI/10.13039/501100011033 is fully acknowledged for the holding of this Conference.

MDPI

Proceeding Paper

Evaluation of a Low-Glucose Gluten-Free Bread Made from Hydrolyzed Cassava Starch and Lupine Flour [†]

Pedro Maldonado-Alvarado * and Vanessa Abad-Quevedo

Department of Food Science and Biotechnology, National Polytechnique School, Quito P.O. Box 17-01-2759, Ecuador; vanessa.abad@epn.edu.ec

* Correspondence: pedro.maldonado@epn.edu.ec

[†] Presented at the V International Conference la ValSe-Food and VIII Symposium Chia-Link, Valencia, Spain, 4–6 October 2023.

Abstract: Currently, there is an increase in diabetes cases and people sensitive to gluten. However, there are few foods in commerce with good quality and that serve to appease, in synergy, these diseases. The aim of this work was to evaluate a low-glucose gluten-free bread made from hydrolyzed cassava starch and lupine flour. The starch of the cassava variety INIAP 651 (CM1335–4 genotype), cultivated in Ecuador, as well as the debittered flour of lupine (*Lupinus mutabilis* Sweet) from Ecuador were used. A cassava starch/water dilution was gelatinized, lyophilized, ground, and sieved. Then, it was hydrolyzed with pancreatic α-amylase prepared at 100 U/g, for 0, 1, 2, and 3 h. In addition, breads were made from gels without the addition of yeast. The gels showed significant differences for the hydrolysis times 0 and 1 h, in hydrolysis level, consistency, cohesiveness, firmness and viscosity level. No significant differences were found in those parameters for 1, 2 and 3 h. In the bread, significant differences were found for 0 and 1 h in specific volume, firmness, springiness, cohesiveness, adhesiveness, and resilience. There were no significant differences between 0 to 3 h of hydrolysis for those parameters. Up to 10% partial substitution of hydrolyzed cassava starch by lupine flour, there were no significant differences in rheological properties.

Keywords: bread; cassava; gluten-free; lupin; modified starch; rheology

Citation: Maldonado-Alvarado, P.; Abad-Quevedo, V. Evaluation of a Low-Glucose Gluten-Free Bread Made from Hydrolyzed Cassava Starch and Lupine Flour. *Biol. Life Sci. Forum* **2023**, *25*, 1. https://doi.org/10.3390/blsf2023025001

Academic Editors: Claudia M. Haros, Loreto Muñoz and María Dolores Ortolá

Published: 27 September 2023

1. Introduction

The mechanism that confers the expansion in sour cassava starch bread, without the use of leavening agents, is given by supramolecular and molecular degradations which come from traditional starch processes of spontaneous fermentation (30 days) and solar drying (12 h), respectively. These treatments are irregular and as a result, bread-making products of different quality are obtained. Previous studies have tried to mimic these processes with treatments that make it possible to control starch granule degradation (addition of organic acids) and/or oxidation at the molecular level (using UV-Vis lamp, sodium hypochlorite, ozone) [1]. However, the effect of degradation at the granular level by α-amylase to achieve a bread-making product with good properties has not been investigated. α-amylase hydrolyzes the alpha-1-4 bonds of starch, depolymerizing it. On the other hand, in the starch granule, the degradations required to achieve this expansion could be minimal, of the order of 1%, according to a single existing reference [2]. This would produce a minimum amount of reducing sugars, ideal for making bread low in simple sugars. However, the minimum level of hydrolysis necessary to achieve adequate breadmaking has not been precisely determined.

Likewise, proteins of vegetal origin such as lupine from Ecuador (*Lupinus mutabilis* Sweet) have been used in order to improve the nutritional content of the bread due to its high protein (40–45%), fiber (25–30%), calcium, etc. In addition, lupin intake helps to increase satiety and reduces energy intake and LDL-cholesterol level in the blood [3].

Also, certain studies have explained that the mechanisms of action of lupine proteins, including a strange protein called gamma conglutin, inhibit the enzyme DPP-4 (dipeptidyl peptidase-4), which favors glucose control in patients with type 2 diabetes; increase glucose uptake in insulin-dependent cells; and they also inhibit gluconeogenesis (glucose production in the liver), as does metformin [4].

The aim of this contribution was to study a gluten-free low-in-free-sugar bread made from modified cassava starch with alpha-amylase and lupine flour.

2. Materials and Methods

For this work, the starch of the cassava variety (Manihot esculenta Crantz) INIAP 651, from the CM1335–4 genotype, cultivated in Manabí-Ecuador, was used. A starch/water (1:20) dispersion was gelled, lyophilized, ground and sieved (106 μm). Then, it was hydrolyzed with pancreatic α-amylase (A 6255, Sigma-Aldrich, St. Louis, MO, USA), prepared at 500 U/g, for 0, 1, 2, and 3 h. Hydrolysates were analyzed for enzymatic digestibility using the Megazyme D-glucose kit (Megazyme International Ireland Ltd., Bray, Ireland) by UV-Vis to determine enzymatic hydrolysis level. For the analysis of rheological parameters, the same gel preparation procedure mentioned above was carried out except for lyophilization. A Perten TVT 6700 texturometer ((Perten Instruments, Hägersten, Sweden) coupled to a probe of 45 mm diameter compression plate and a 120 mL container was used to analyze the hydrolyzed gels. The 24-01.02 Curdled Consistency-Back extrusion test from TexCal 5 software of the texturometer was used to analyze the rheological properties of the obtained gel and bread. On the other hand, breads without the addition of yeast made from the described gels were elaborated, and the specific volume and textural properties were evaluated. In addition, a bread partial substitution of modified cassava starch with lupin flour (0, 5, 10, 15 and 20%) was performed to determine the loaf volume of bread. To determine statistically significant differences, an ANOVA (analysis of variance) followed by LSD (Fisher´s least significant differences) test were performed ($p < 5\%$; n = 3).

3. Results and Discussion

Table 1 shows results from alpha-amylase cassava starch gel and bread. The gels showed significant differences for the hydrolysis times 0 and 1 h, in hydrolysis level (10 and 63% w/w), consistency (158 and 350 gf.mm), cohesiveness (−26.6 and 0 gf), firmness (32.2 and 14.7 gf) and viscosity level (158 and 0 gf.mm). No significant differences were found in those parameters for 1, 2 and 3 h. In the bread, significant differences were found for 0 and 1 h in specific volume (1.93–2.47 mL/g) [1], firmness (4106–4774 gf), springiness (80–93%), cohesiveness (0.83–0.97), adhesiveness (−60.5−−114.3 gf.mm) and resilience (0.63–0.925). There were no significant differences between 0 to 3h of hydrolysis for those parameters. These results suggest that only 1 h of amylase hydrolysis, or perhaps less, is enough to significantly modify the level of hydrolysis and the rheological properties of gelatinized starch. However, when evaluating the finished product with the different hydrolysis times, there are no differences between the treatments. An additional modification with UV-Vis to the hydrolyzed starch could cause positive degradations at the molecular level and contribute to improving the functional properties of the bread that can be correlated with the other results obtained.

On the other hand, specific volumes of bread with partial substitution of modified cassava starch with lupin flour at 0, 5, 10, 15 and 20% were, respectively, 2.47; 2.39; 2.40; 2.07; and 1.91 mL/g. There were no significant differences for 0, 5 and 10% of bread partial substitution in loaf volume and textural properties (resilience, springiness, adhesiveness, firmness, and cohesiveness) [5].

Table 1. Results from gel and bread made from alpha-amylase cassava starch.

Parameter/Time	0 min	60 min	120 min	180 min
Gel-amylase cassava starch				
Consistency (gf mm)	158.67 ± 9.29 [a]	350.01 ± 9.54 [b]	341.67 ± 25.54 [b]	329.00 ± 8.00 [b]
Cohesiveness (g)	−26.33 ± 0.58 [a]	0.00 ± 0.00 [b]	0.00 ± 0.00 [b]	0.00 ± 0.00 [b]
Firmness (g)	32.33 ± 0.58 [b]	14.67 ± 0.58 [a]	14.67 ± 0.58 [a]	14.00 ± 0.00 [a]
Enzymatic hydrolysis (%)	10.01 ± 5.51 [a]	63.09 ± 0.52 [b]	56.21 ± 7.45 [b]	60.70 ± 1.87 [b]
Bread-amylase cassava starch				
Loaf volume (g/mL)	0.91 ± 0.04 [a]	0.87 ± 0.03 [a]	0.91 ± 0.05 [a]	0.94 ± 0.00 [a]
Resilience	0.63 ± 0.00 [a]	0.70 ± 0.10 [a]	0.725 ± 0.00 [a]	0.69 ± 0.00 [a]
Springiness	0.90 ± 0.00 [a]	0.93 ± 0.00 [a]	0.91 ± 0.00 [a]	0.90 ± 0.00 [a]
Adhesiveness (g f mm)	−88 ± 17 [a]	77 ± 15 [a]	−60 ± 28 [a]	−114 ± 28 [a]
Firmness (g)	477 ± 23 [a]	410 ± 13 [a]	455 ± 20 [a]	436 ± 26 [a]
Cohesiveness	0.83 ± 0.00 [a]	0.90 ± 0.00 [a]	0.90 ± 0.10 [a]	0.88 ± 0.10 [a]

[a,b] Different letters, within row, indicate significant differences at $p < 0.05$ (Fisher). n = 3.

4. Conclusions

Gel and bread made from cassava starch modified with α-amylase did not change rheological properties from 1 h of amylase hydrolysis. It would be necessary to explore different hydrolysis times between 0 and 1 h to determine changes in properties. Up to 10% partial substitution of hydrolyzed cassava starch by lupine flour, there are significant differences in breadmaking properties.

Author Contributions: Conceptualization, P.M.-A.; Formal analysis, P.M.-A. and V.A.-Q.; Methodology, V.A.-Q.; Supervision, P.M.-A.; Validation, P.M.-A.; Writing—original draft, P.M.-A.; Writing—review and editing, P.M.-A. All authors have read and agreed to the published version of the manuscript.

Funding: This work was supported by grant Ia ValSe-Food (119RT0567) and financially supported by EPN through the project PIS 21-04 of Ecuador.

Institutional Review Board Statement: Not applicable.

Informed Consent Statement: Not applicable.

Data Availability Statement: The data analyzed in this study are available from the authors upon reasonable request.

Conflicts of Interest: The authors declare no conflict of interest.

References

1. Maldonado-Alvarado, P.; Grosmaire, L.; Dufour, D.; Giraldo-Toro, A.; Sánchez, T.; Calle, F.; Moreno-Santander, M.A.; Ceballos, H.; Delarbre, J.-L.; Tran, T. Combined Effect of Fermentation, Sun-Drying and Genotype on Breadmaking Ability of Sour Cassava Starch. *Carbohydr. Polym.* **2013**, *98*, 1137–1146. [CrossRef] [PubMed]
2. Camargo, C.; Colonna, P.; Buleon, A.; Richard-Molard, D. Functional Properties of Sour Cassava (Manihot Utilissima) Starch: Polvilho Azedo. *J. Sci. Food Agric.* **1988**, *45*, 273–289. [CrossRef]
3. Lee, Y.P.; Mori, T.A.; Sipsas, S.; Barden, A.; Puddey, I.B.; Burke, V.; Hall, R.S.; Hodgson, J.M. Lupin-Enriched Bread Increases Satiety and Reduces Energy Intake Acutely. *Am. J. Clin. Nutr.* **2006**, *84*, 975–980. [CrossRef] [PubMed]
4. Muñoz, E.B.; Luna-Vital, D.A.; Fornasini, M.; Baldeón, M.E.; Gonzalez de Mejia, E. Gamma-Conglutin Peptides from Andean Lupin Legume (Lupinus Mutabilis Sweet) Enhanced Glucose Uptake and Reduced Gluconeogenesis in Vitro. *J. Funct. Foods* **2018**, *45*, 339–347. [CrossRef]
5. Rosell, C.M.; Cortez, G.; Repo-Carrasco, R. Breadmaking Use of Andean Crops Quinoa, Kañiwa, Kiwicha, and Tarwi. *Cereal Chem. J.* **2009**, *86*, 386–392. [CrossRef]

Proceeding Paper

Development of Powdered Beverage with Cushuro (*Nostoc commune*) Concentrated Protein and Quinoa (*Chenopodium quinoa*) †

Nancy Chasquibol *, Axel Sotelo and Rafael Alarcón

Ingeniería Industrial, Grupo de Investigación en Alimentos Funcionales, Instituto de Investigación Científica, Universidad de Lima, Av. Javier Prado Este, 4600, Fundo Monterrico Chico, Surco, Lima 15023, Peru; aasotelo@ulima.edu.pe (A.S.); ralarcor@ulima.edu.pe (R.A.)

* Correspondence: nchasquibol@ulima.edu.pe

† Presented at the V International Conference la ValSe-Food and VIII Symposium Chia-Link, Valencia, Spain, 4–6 October 2023.

Abstract: The trend in protein consumption is towards plant-based foods, for this reason a powdered beverage with a balanced profile of essential amino acids was developed, combining two different sources of protein, cushuro (*Nostoc commune*) and quinoa (*Chenopodium quinoa*). The cushuro is an ancient Andean cyanobacteria that grows over 3000 and 4000 m above sea level, and quinoa is an important Andean seed, and they are part of the basic diet of the Andean Peruvian population. The cushuro obtained from the region of Ancash-Peru was dried (60 °C, 18 h) and the flour was dissolved in 0.1 N HCl at 90 °C for 20 min to obtain a protein concentrate with 6.79% moisture, 0.16% fat, 71.03% protein, and good balance of essential amino acids. Quinoa flour presented 8.52% moisture, 4.84% fat, 12.90% protein, and essential amino acids such as phenylalanine, and lysine. Quinoa flour (65.45%), cushuro concentrated protein (4.76%), cocoa (17.85%), panela (11.90%), and stevia (0.05%), were mixed to obtain the powdered beverage. The results showed that powdered beverage presented an adequate balance of amino acids according to FAO/WHO, with 14.36% proteins, 72.53% carbohydrate, 5.19% fat, 716.6 mg/kg potassium, 319.8 mg/kg phosphorus, 139.2 mg/kg magnesium, 82.69 mg/100 g vitamin C, 1.49 µg/100 g vitamin B12, heavy metals below the detection evel (<0.050 mg/kg). The powdered beverage complied with the Peruvian legislation of the Healthy Law No. 30021 and with the microbiological requirements.

Keywords: concentrated protein; cushuro; *Nostoc commune*; quinoa; beverage

Citation: Chasquibol, N.; Sotelo, A.; Alarcón, R. Development of Powdered Beverage with Cushuro (*Nostoc commune*) Concentrated Protein and Quinoa (*Chenopodium quinoa*). *Biol. Life Sci. Forum* **2023**, *25*, 2. https://doi.org/10.3390/blsf2023025002

Academic Editors: Claudia M. Haros, Loreto Muñoz and María Dolores Ortolá

Published: 28 September 2023

1. Introduction

Cushuro belongs to the genus *Nostoc*, a type of algae (cyanobacteria), when they are hydrated, they can form spheres of 10 to 25 mm in diameter, like grapes. Cushuro is found in the Andean foothills over 3000 m above sea level. Its characteristic color is bluish green and is part of the basic diet of the Andean Peruvian population. Cushuro is also known as "murmunta", "llullucha" or "llayta". The *Nostoc commune* Vauch. form colonies that in their natural environment resist dryness and can easily restore their metabolism after being rehydrated because cells produce large amounts of protective polysaccharides. Cushuro is source of nutrients such as protein, calcium, and iron as well as bioactive compounds [1].

The quinoa (*Chenopodium quinoa*) seed is a pseudocereal with high protein content and excellent amino acid composition, domesticated 5000 years ago in the Andean region, due to its high nutritional value, and bioactive properties including antioxidant, hypolipidemic, immunomodulatory, weight-regulating, hypoglycemic, hypotensive, probiotic, antitumor, and hormone-regulating effects [2].

Functional powdered beverages with plant-based protein have been developed as a strategy to get a better nutritional balance, however, few plant species have been used in the industry for this purpose [3]. The present research focuses on the obtention and characterization of a powdered beverage of cushuro (*Nostoc commune*) concentrated protein and quinoa (*Chenopodium quinoa*) flour, with a balanced profile of essential amino acids.

2. Materials and Methods

2.1. Raw Material

The cushuro was obtained from Cotaparaco distric, Recuay province, Ancash region-Peru and quinoa was obtained from a local market from the local market in Lima city. The cushuro was washed, dried at 60 °C for 12 h in an infrared dehydrate (IRC D18, Irconfort, Sevilla, Spain), then ground in the food shredder (Grindomix GM200, Restch, Haan, Germany) in the Functional Foods Laboratory from the Universidad de Lima-Peru to obtain cushuro flour. Quinoa was dried at 40 °C and ground in the food shredder. All the samples were stored in aluminized bags at room temperature until use.

2.2. Extraction of Cushuro Concentrated Protein (CCP)

The cushuro flour was pretreated with HCl 0.1 N for 20 min (64.44 mL/g, mL solvent/g sample) by ultrasound-assisted microwave (CW-2000, Nade, Shanghai, China) extraction (500 W, 50 W/60 Hz). The result of the extraction was filtered under vacuum with muslin cloth. The solid residue (CCP) was washed with water to remove excess acid and dehydrated for 12 h in infrared dryer at 60 °C (IRC D18, Irconfort, Sevilla, Spain). The CCP solid was ground (Grindomix GM200, Restch, Haan, Germany) and stored in aluminized bags for later use.

2.3. Development of Powdered Beverage (PB)

The powdered beverage was developed by mixing quinoa flour (27.5 g), cushuro protein flour (2 g), cocoa (7.5 g), cane sugar (5 g) and stevia (0.02 g) in a planetary mixer (FPSTSMPL1-053, Oster, Guangdong, China). The mixture was packaged in aluminized bags with a net weight of 42 g per unit. The formulation was made to obtain a PB with a protein percentage greater than 14%.

2.4. Proximal Composición

The proximal composition was carried out according to official methods. The moisture content was determined at 110 °C to constant weight. The total protein content was determined as % nitrogen × 6.25 using a Kjeldahl analyzer (UDK 139, VELP, Usmate Velate, Italy). The ash content was determined by incineration at 550 °C for 72 h, in a muffle furnace. The fat content was determined with hexane for 9 h.

2.5. Determination of Mineral Content

The concentrations of Ca, Cu, Fe, K, Mg, Na, P, Zn and heavy metals such as As, Pb, Hg and Cd in the powdered beverage were determined by atomic absorption spectrophotometry (Nexion 350x, Perkin Elmer, MA, USA) according to AOAC Official Method 2015.01 (2015).

2.6. Determination of Vitamins C and Vitamin B12

The concentrations of Vitamin C and Vitamin B12 were determined by AOAC 985.33 (2019) and AOAC 952.20 (2019), respectively.

2.7. Determination of Solubility, Bulk and Compacted Density and Hausner Index

Solubility, bulk and compacted density and Hausner Index were determined according to Chasquibol et al. [4].

2.8. Determination of Amino Acid Profile

The amino acid profile was estimated according to Chasquibol et al. [5], and it was quantified by HPLC (ARC, Waters, Milford, CT, USA) with an inner diameter of 150 mm × 3.9 mm. column C18 reverse phase.

2.9. Determination of Na, Total Sugar, Saturated Fat, and Trans Fat

The concentrations of Na, total sugar, saturated fat, and *trans* fat were determined by AOAC 969.23 (2019), AOAC 968.28 (2019), ISO 12966-1 (2014) and ISO 12966-1 (2014) respectively.

2.10. Microbiological Analysis

The powdered beverage was analyzed microbiologically to stablish the microbiological criteria of sanitary quality and safety for food and beverages for human consumption.

2.11. Statistical Analysis

Results were expressed as mean ± standard deviation. All measurements were determined in duplicate or triplicate. Analysis of variance (ANOVA), which was used to analyze data acquired at a 95% significance level with Minitab 19.0 Software (Minitab Inc., State College, Palo Alto, CA, USA).

3. Results and Discussions

Table 1 shows the physicochemical characterization of PB, CCP, and quinoa flour (QF). The powdered beverage had a high protein content (14.36 ± 0.33%). According to FAO, 42 g of PB rehydrated in 400 mL of water, would cover 20% of the protein recommended daily for children under 5 years old with an average weight of 20 kg. Also, according to Codex Alimentarius CAC/GL 23-1997, the PB could be declared as a high-protein food for children under 5 years of age. Besides, the carbohydrate (72.53 ± 0.36%), ash (3.29 ± 1.15%) and fat content (5.19 ± 0.01%) increased, but moisture (4.63 ± 0.11%) content decreased.

Table 1. Proximal composition of cushuro concentrated protein (CCP), quinoa flour (QF) and powdered beverage (PB).

Determination	CCP	QF	PB
Moisture (g/100 g)	6.79 ± 0.28	8.52 ± 0.11	4.63 ± 0.11
Ash (g/100 g)	0.94 ± 0.00	2.54 ± 0.08	3.29 ± 1.15
Fat (g/100 g)	0.16 ± 0.03	4.84 ± 0.02	5.19 ± 0.01
Carbohydrates (g/100 g)	21.41 ± 0.10	71.2 ± 0.21	72.53 ± 0.36
Protein (g/100 g)	71.03 ± 0.73	12.9 ± 0.15	14.36 ± 0.33

Results are expressed as mean ± SD (n = 3).

The composition of amino acids profile is summarized in Table 2. The predominant amino acids of PB were aspartic acid + asparagine (120.5 ± 0.58 mg/g protein), glutamic acid + glutamine (140.7 ± 10.29 mg/g protein), serine (55.79 ± 1.63 mg/g protein), glycine (56.66 ± 2.54 mg/g protein), threonine (71.19 ± 2.12 mg/g protein), arginine (63.36 ± 4.36 mg/g protein), alanine (63.30 ± 1.88 mg/g protein), valine (60.39 ± 1.32 mg/g protein) and leucine (64.60 ± 3.95 mg/g protein). The amino acid composition of PB showed a balanced profile, according to the FAO/WHO recommendations. Limited amino acids (leucine 58.95 ± 2.64 mg/g protein, and tryptophan 6.00 ± 0.02 mg/g protein) in quinoa were improved in the CCP (leucine 64.00 ± 0.00 mg/g protein, and tryptophan 8.60 ± 0.00 mg/g protein). Thus, the powdered beverage showed higher content of specific amino acids (leucine 64.6 ± 3.95 mg/g protein, and tryptophan 27.02 ± 4.37 mg/g protein). Bonke et al. [6] showed that beverages formulated with oat, lentil and pea were rich in amino acids, except for tryptophan. In this research, the powdered beverage was rich in tryptophan, three times above the recommendations of the FAO/WHO.

Table 2. Amino acids profile of cushuro concentrated protein (CCP), quinoa flour QF) and powdered beverage (PB).

Amino Acids	CCP (mg Amino Acid/g Protein)	QF (mg Amino Acid/g Protein)	PB (mg Amino Acid/g Protein)	OMS
Aspartic acid + Asparagine	149.80 ± 0.80	70.27 ± 16.32	120.5 ± 0.58	
Glutamic acid + Glutamine	80.50 ± 3.20	123.83 ± 14.55	140.7 ± 10.29	
Serine	59.20 ± 0.20	43.62 ± 1.98	55.79 ± 1.63	
Histidine	2.20 ± 0.10	27.43 ± 0.11	16.6 ± 0.87	15
Glycine	57.30 ± 0.40	51.02 ± 2.21	56.66 ± 2.54	
Threonine	121.90 ± 0.40	37.06 ± 1.75	71.19 ± 2.12	23
Arginine	44.80 ± 0.40	73.4 ± 2.08	63.36 ± 4.36	
Alanine	87.30 ± 0.40	39.88 ± 3.89	63.6 ± 1.88	
Proline	42.90 ± 2.30	22.1 ± 8.9	22.59 ± 1.07	
Tyrosine [b]	17.80 ± 0.10	22.8 ± 1.95	25.03 ± 1.93	38 [b]
Valine	103.80 ± 0.10	42.34 ± 1.95	60.39 ± 1.32	39
Methionine [a]	1.50 ± 0.10	1.37 ± 0.28	8.44 ± 0.86	22 [a]
Cysteine [a]	1.10 ± 0.00	4.27 ± 0.21	4.69 ± 0.62	
Isoleucine	57.80 ± 0.00	34.81 ± 1.62	43.57 ± 1.5	30
Tryptophan	8.60 ± 0.00	6.00 ± 0.02	27.02 ± 4.37	6
Leucine	64.00 ± 0.00	58.95 ± 2.64	64.6 ± 3.95	59
Phenylalanine [b]	69.30 ± 0.12	37.18 ± 1.09	48.4 ± 1.57	
Lysine	30.10 ± 0.10	52.1 ± 1.63	42.17 ± 3.25	45

Results are expressed as mean ± SD (n = 3). Note: According to the Food and Agriculture Organisation, 'Assessment of the quality of dietary proteins in human nutrition' (2013). [a]: Met + Cys; [b]: Phe + Tyr.

The PB had a compacted density of 0.935 g/mL and a Hausner index of 1.03. These values indicate that the powdered beverage has little fluidity and would occupy a smaller volume in its packaging. The solubility values obtained are lower (28.08%) by the higher fiber contain the quinoa, extrusion process would improve the solubility of the beverage.

The PB contained important quantity of minerals (Table 3): calcium (130.6 ± 10.40 mg/kg), potassium (716.6 ± 57.28 mg/kg), magnesium (139.2 ± 11.12 mg/kg), sodium (62.4 ± 4.96 mg/kg), phosphorus (319.8 ± 25.52 mg/kg), copper (0.8 ± 0.06 mg/kg), iron (4.0 ± 0.32 mg/kg) and zinc (2.8 ± 0.22 mg/kg). According to the daily requirements for children under 5 years of the Institute of Medicine (US) Standing Committee on the Scientific Evaluation of Dietary Reference Intakes [7] and National Institute of Health [8], the PB would cover 6.8% about requirement daily of calcium, 26.86% phosphorus, 44.97% magnesium, 13.08% potassium and 16.8% iron. So, the PB developed could contribute to the reduction of malnutrition and childhood anemia in our country. The presence of heavy metals (Hg, As, Cd and Pb) was not detected in the PB (<0.050 mg /kg).

Table 3. Minerals, heavy metals and vitamins content of powdered beverage (PB).

Minerals (mg/kg)		Heavy Metal (mg/kg)		Vitamins	
Calcium	130.6 ± 10.40	Arsenic	<0.050 ± 0.001	Vitamin C (mg/100 g)	82.69 ± 0.10
Copper	0.8 ± 0.06	Cadmium	<0.050 ± 0.001	Vitamin B12 (µg/100 g)	1.49 ± 0. 01
Iron	4.0 ± 0.32	Mercury	<0.050 ± 0.001		
Potassium	716.6 ± 57.28	Lead	<0.050 ± 0.001		
Magnesium	139.2 ± 11.12				
Sodium	62.4 ± 4.96				
Phosphorus	319.8 ± 25.52				
Zinc	2.8 ± 0.22				

Results are expressed as mean ± SD (n = 3).

The PB had higher vitamin C (82.69 mg/100 g) than commercial brands in Peru and the vitamin B12 (1.49 µg/100 g) content could cover 51.1% of the daily requirement according to the National Institute of Health.

The content of sodium, total sugar, saturated fat, and trans fat of the PB is shown in Table 4. The PB have a low content in sodium (46.02 ± 3.68 mg/100 g), total sugar (9.33 ± 0.72 g/100 g), saturated fat (1.39 ± 0.01 g/100 g) and trans-fat ($<0.01 \pm 0.001$ g/100 g fat), below the limits permissible established in Peruvian legislation.

Table 4. Sodium, total sugars, saturated fat and *trans* fats content of powder beverage (PB).

Determination	PB	Maximum According to Standard *
Sodium mg/100 g	46.02 ± 3.68	400 mg/100 g
Total sugars (g/100 g)	9.33 ± 0.72	10 g/100 g
Saturated fat (g/100 g)	1.39 ± 0.01	4 g/100 g
Trans fats (g /100 g fat)	$<0.01 \pm 0.001$	5 g/100 g

Results are expressed as mean $\pm$ SD (n = 3). * Law No. 30021 Law on the Promotion of Healthy Eating for Children and Adolescents, regulated in Supreme Decree No. 017-2017-SA-Peru.

The microbiological analysis of powder beverage (Table 5) showed that the PB complied with the microbiological standard according to the limits permissible, established in Peruvian legislation of sanitary quality and safety for food and beverages for human consumption.

Table 5. Microbiological analysis of powder beverage (PB).

Microorganism	PB	Maximum According to Standard *
Staphylococcus aureus (CFU/100 g)	$<10 \pm 0.01$	10^2
Coliformes (CFU/100 g)	$<10 \pm 0.01$	10^2
Salmonella sp. (presence/absence)	Absence	Absence
Bacillus cereus (CFU/100 g)	$<100 \pm 0.10$	10^3

Results are expressed as mean $\pm$ SD (n = 3). * Ministerial Resolution 591-2008-MINSA-Peru. Sanitary standard that establishes the microbiological criteria of sanitary quality and safety for food and beverages for human consumption. CFU: Colony Forming Units.

4. Conclusions

The results showed that the PB of cushuro and quinoa had a balanced profile of essential amino acids according to FAO/WHO, being the majority aspartic acid + asparagine, glutamic acid + glutamine, serine, glycine, threonine, arginine, alanine, valine, and leucine. In addition, PB had important quantity of minerals such as potassium, phosphorus, and iron. The PB accomplishes the microbiology sanitary standards of the Peruvian law and the healthy standards legislation for children Thus, the powdered beverage could be an important supplement for children to promote a healthy nutrition.

Author Contributions: All authors have contributed equally to this manuscript. Conceptualization, N.C., A.S. and R.A.; Methodology, N.C., A.S. and R.A.; Investigation and Data analysis, N.C., A.S. and R.A.; Writing—original draft preparation, N.C., A.S. and R.A.; Writing—review and editing, N.C. All authors have read and agreed to the published version of the manuscript.

Funding: This research was funded by Project N°381-2020-Programa Nacional de Investigación en Pesca y Acuicultura (PNIPA), belonging to "The Ministry of Production-Peru" and Universidad de Lima-Peru.

Institutional Review Board Statement: Not applicable.

Informed Consent Statement: Not applicable.

Data Availability Statement: Data is contained within the manuscript.

Acknowledgments: The authors thank the La ValSe-Food (119RT0567), Programa Nacional de Investigación en Pesca y Acuicultura (PNIPA) and Universidad de Lima-Peru.

Conflicts of Interest: The authors declare no conflict of interest.

References

1. Méndez, A.S.; Victoriano, P.; Soto, G.H.; Zambrano, C.W.; Marín, M.O.; Zapata, R.J.; Maquera, M.M.; Fernandez, H.R.; González, A.J.; Zuffo, A.; et al. Physicochemical Evaluation of Cushuro (*Nostoc sphaericum* Vaucher ex Bornet & Flahault) in the Region of Moquegua for Food Purposes. *Foods* **2023**, *12*, 1939. [CrossRef]
2. Ng, C.Y.; Wang, M. The functional ingredients of quinoa (*Chenopodium quinoa*) and Physiological effects of consuming quinoa: A review. *Food Front.* **2021**, *2*, 329–356. [CrossRef]
3. Arbach, C.T.; Alves, I.A.; Serafini, M.R. Recent patent applications in beverages enriched with plant proteins. *Sci. Food* **2021**, *5*, 28. [CrossRef] [PubMed]
4. Chasquibol, N.; Alarcón, R.; Gonzales, B.F.; Sotelo, A.; Landoni, L.; Gallardo, G.; García, B.; Pérez-Camino, M.C. Design of Functional Powdered Beverages Containing Co-Microcapsules of Sacha Inchi *P. huayllabambana* Oil and Antioxidant Extracts of Camu Camu and Mango Skins. *Antioxidants* **2022**, *11*, 1420. [CrossRef] [PubMed]
5. Chasquibol, N.; Gonzales, B.F.; Alarcón, R.; Sotelo, A.; Márquez-López, J.C.; Rodríguez-Martin, N.M.; del Carmen Millán-Linares, M.; Millán, F.; Pedroche, J. Optimisation and Characterisation of the Protein Hydrolysate of Scallops (*Argopecten purpuratus*) Visceral By-Products. *Foods* **2023**, *12*, 2003. [CrossRef] [PubMed]
6. Walther, B.; Guggisberg, D.; Badertscher, R.; Egger, L.; Portmann, R.; Dubois, S.; Haldimann, M.; Kopf-Bolanz, K.A.; Rhyn, P.; Zoller, O.; et al. Comparison of nutritional composition between plant-based drinks and cow's milk. *Front. Nutr.* **2022**, *9*, 988707. [CrossRef] [PubMed]
7. Institute of Medicine (US) Standing Committee on the Scientific Evaluation of Dietary Reference Intakes. *Dietary Reference Intakes for Calcium, Phosphorus, Magnesium, Vitamin D, and Fluoride*; National Academies Press: Washington, DC, USA, 1997.
8. National Institutes of Health. Hierro. Hoja Informativa para Consumidores. Available online: https://ods.od.nih.gov/factsheets/Iron-DatosEnEspanol/ (accessed on 1 August 2023).

biology and life sciences forum

Proceeding Paper

Obtaining Quinoa Germ via Wet Milling and Extracting Its Oil via Cold Pressing [†]

Ana Ribera-Castelló and Claudia Monika Haros *

Instituto de Agroquímica y Tecnología de Alimentos (IATA), The Spanish National Research Council (CSIC), Av. Agustín Escardino 7, Parque Científico, 46980 Paterna, Valencia, Spain; a.ribera@iata.csic.es

* Correspondence: cmharos@iata.csic.es

[†] Presented at the V International Conference la ValSe-Food and VIII Symposium Chia-Link, Valencia, Spain, 4–6 October 2023.

Abstract: Wet milling is a fractionation process widely used in the corn industry, which allows the separation of its main chemical components (starch, proteins, fiber and lipids) with high efficiency and purity compared to dry milling. The first stage of this process consists of maceration; after softening the grain, the actual milling is carried out, and the germ is separated by flotation because of its high lipid content. The chemical composition of pseudocereals is similar to that of cereals, hence their name, so they could be processed in the same way. In this way, the traditional corn wet milling process was adapted to quinoa. The objective of this work is to isolate the germ of red Bolivian Royal quinoa using wet milling, and evaluate its efficiency and physicochemical characteristics due to its large size and nutrient concentration. By cold pressing the red quinoa germ, crude oil was obtained and characterized in terms of: Acid Index, Iodine Index, Saponification Index, K Index, Refractive Index (20 °C) and fatty acid composition, determined by gas chromatography coupled to a mass detector (GC-MS). This profile was compared with the fatty acid profile of the solvent-extracted quinoa oil, and it was observed that there were no significant differences between the two oil samples. In addition, the sample obtained via cold pressing showed similar characteristics to corn oil, except for a higher Saponification Index and proportion of linolenic acid (omega-3).

Keywords: fatty acid profile; red Bolivian Royal quinoa; quinoa germ; quinoa oil; quinoa wet milling

Citation: Ribera-Castelló, A.; Haros, C.M. Obtaining Quinoa Germ via Wet Milling and Extracting Its Oil via Cold Pressing. *Biol. Life Sci. Forum* **2023**, *25*, 3. https://doi.org/10.3390/blsf2023025003

Academic Editors: Loreto Muñoz and María Dolores Ortolá

Published: 28 September 2023

1. Introduction

Quinoa grain is a food of high nutritional quality and a source of bioactive compounds, which are concentrated in specific anatomical parts of the grain [1]. The study of quinoa grain fractionation could be approached by adapting the processes currently used in the primary cereal processing industry, specifically dry milling of wheat or wet milling of corn [2] (Figure 1). Wet milling is a fractionation process that allows the high recovery and purity of the chemical components of the grain (starch, proteins, fiber, and lipids) to be obtained compared to dry milling [3].

The objective of this work is to isolate the germ of red Royal Bolivian quinoa via wet milling, and evaluate its efficiency and physicochemical characteristics. Subsequently, the oil was obtained through cold pressing of the quinoa germ, and its fatty acid profile was compared with the oil obtained through organic solvent extraction.

WET MILLING - Germ Fraction

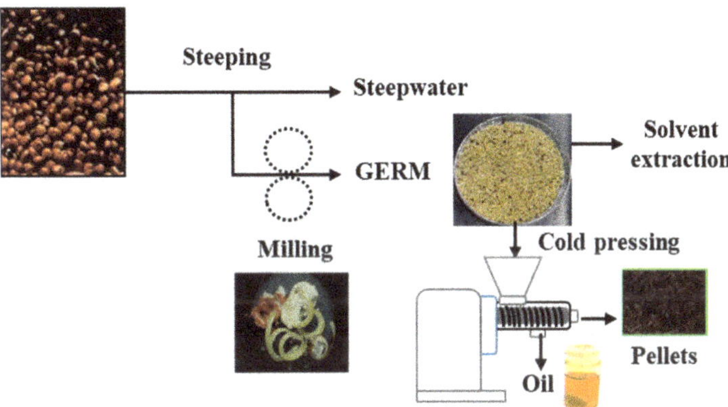

Figure 1. Flow sheet of quinoa wet milling and its oil extraction.

2. Materials and Methods

The material used in this research was organic red Royal quinoa from Uyuni, Bolivia.

2.1. Obtaining and Characterization of Germ

After quinoa wet milling, the germ was separated through flotation according to Ballester et al. [2]. The efficiency of germ fraction was evaluated according to the following equation:

$$\text{Efficiency \%} = \frac{\text{Dry weight of germ fraction}}{\text{Initial dry weight of seeds}} \times 100$$

The proximate composition of the grain and germ of the red Royal quinoa was characterized in terms of moisture, lipids, starch, protein and ash according to Miranda et al. [4].

2.2. Characterization of Quinoa Oil

The extraction procedure was conducted by cold pressing the quinoa germ in an oil Press machine stainless steel (Yameijia A-2288, Zhongshan, China). Subsequently, a liquid–solid extraction with hexane was carried out in a lipid extractor Randall (Velp Scientifica, SER158 Solvent Autoextractor, Usmate, Italy).

The characterization of the oil was carried out using the following parameters: Acidity Index (mg KOH needed to neutralize 1 g of sample), Saponification Index (mg KOH/g oil) and Refractive Index (20 °C) by means of the regulations BOE-A-1977-16116 [5]; Iodine Index (mg I_2/100 g) UNE-EN 14111:2022 [6] and K_{232} Index (Absorbance measured at wavelength 232 nm) UNE-EN-ISO 3656:2011/A1:2017 [7].

In order to obtain the fatty acid profile, the lipids were transesterified to convert them into fatty acid methyl esters [8]. The fatty acid profile was determined by gas chromatography coupled to a mass detector (GC-MS) [9].

2.3. Statistical Analysis

Multiple ANOVA and Fisher's least significant difference (LDS) tests were applied to establish statistically significant differences. Statistical analyses were performed with the software Statgraphics Centurion XVI and the significance level was established at $p < 0.05$.

3. Results

The efficiency of extraction of the red Royal quinoa germ fraction by wet milling was higher than that obtained in corn according to Eckoff et al. [10] (Figure 2A). The fractionation of quinoa resulted in 209% more protein and 84% more lipids compared to the original quinoa grain from which it originated (Figure 2B).

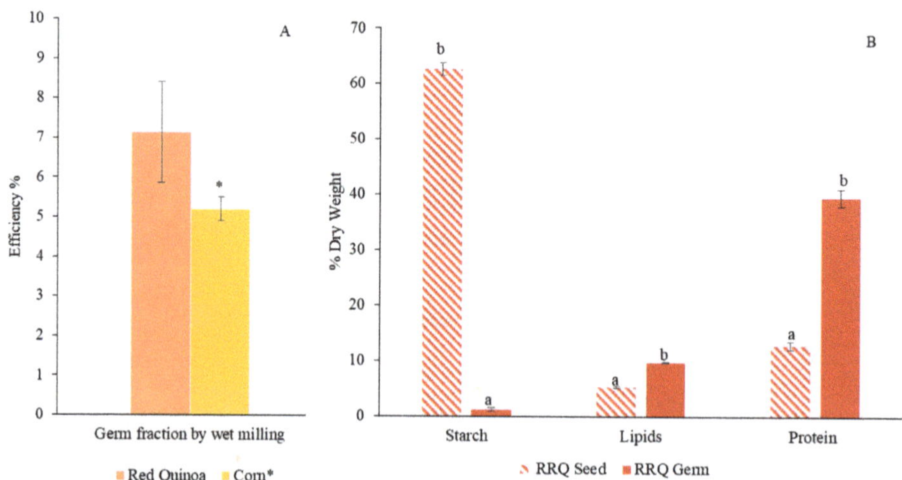

Figure 2. (**A**). Extraction efficiency of the red quinoa germ fraction compared to corn germ fraction obtained by wet milling. * Values retrieved from "A 100-g Laboratory Corn Wet-Milling Procedure" by Eckhoff et al. [10]. (**B**). Red quinoa proximate composition (striped bars) compared to the proximate composition of germ fraction (filled bars). RRQ: Red Royal quinoa; Mean ± SD, n = 3. Bars of the same parameter followed by the same letter indicate that there are no statistically significant differences at a 95% confidence level.

The characterization of red Royal quinoa oil obtained by cold pressing of the germ fraction is shown in Table 1. In addition, the results were compared with corn oil [11]. Both oils showed similar characteristics, except for the Saponification Index, which was higher in quinoa oil than in corn oil, 187–196 mg KOH/g [11].

Table 1. Chemical and physical characteristics of crude quinoa oil obtained by cold pressing.

Parameter	Units	Quinoa Oil
Acid Index	%	14.30 ± 0.08
Iodine Index	mg I_2/100 g	122.2 ± 0.9
Saponification Index	mg KOH/g	273.9 ± 0.8
K_{232} Index	-	0.601 ± 0.164
Refractive Index (20 °C)	-	1.481 ± 0.001

Mean ± SD, n = 3.

We also obtained quinoa oil using solvent extraction. Comparing the fatty acid profile, no significant differences were observed between the red Royal quinoa oil obtained via cold pressing and solvent extraction (Figure 3). As for essential acids, linoleic acid was present in a similar proportion to corn oil, while linolenic acid was present in a higher proportion (Figure 3) [11].

Biol. Life Sci. Forum **2023**, *25*, 3

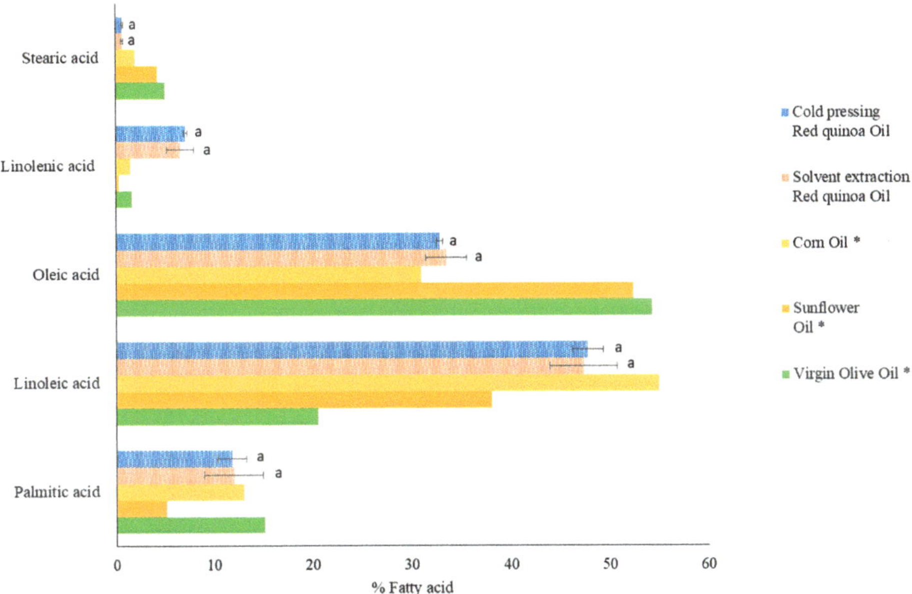

Figure 3. Fatty acid profile of quinoa oil of the germ fraction extracted by cold pressing and solvent extraction compared to the fatty acid profile of the most frequently used edible oils. Mean ± SD, n = 3. Bars of the same parameter followed by the same letter indicate that there are no statistically significant differences at a 95% confidence level. * Values retrieved from BOE-A-1983-5543 [11].

4. Conclusions

The wet milling process of quinoa allows the separation of the germ fraction, which leads to the enrichment of fractions rich in protein and food-grade oil, as is the case with cereals such as corn.

The characteristics of quinoa oil are similar regardless of the extraction process (cold pressing or solvent extraction).

Since quinoa oil has a higher concentration of linolenic acid than other edible oils, its use in food could help reverse the nutritional imbalance in terms of the omega-6/omega-3 ratio in Western diets.

Author Contributions: Conceptualization, C.M.H.; Funding acquisition, C.M.H.; Investigation, A.R.-C. and C.M.H.; Methodology, A.R.-C. and C.M.H.; Project administration, C.M.H.; SupervIsion, C.M.H.; Writing-original draft, A.R.-C. and C.M.H.; Writing-review and editing, C.M.H. All authors have read and agreed to the published version of the manuscript.

Funding: This work was supported by grant Ia ValSe-Food-CYTED (119RT0567), Food4ImNut Food4ImNut (PID2019-107650RB-C21 funded by MCIN/AEI/10.13039/501100011033.

Institutional Review Board Statement: Not applicable.

Informed Consent Statement: Not applicable.

Data Availability Statement: The data presented in this study are available on request from the corresponding author. The data are not publicly available due to their privacy.

Acknowledgments: The contract of Ana Ribera-Castelló from INVESTIGO Program within the framework of the Recovery, Transformation and Resilience Plan of Comunitat Valenciana, Spain.

Conflicts of Interest: The authors declare no conflict of interest.

References

1. Reguera, M.; Haros, C.M. Structure and composition of kernels. In *Pseudocereals: Chemistry and Technology*; Wiley: Chichester, UK, 2017; pp. 28–48. [CrossRef]
2. Ballester, J.; Gil, J.V.; Fernández, M.T.; Haros, C.M. Quinoa wet-milling: Effect of steeping conditions on starch recovery and quality. *Food Hydrocoll.* **2019**, *89*, 837–843. [CrossRef]
3. Haros, C.M.; Wronkowska, M. Pseudocereal dry and wet milling: Processes, products and applications. In *Pseudocereals: Chemistry and Technology*; Wiley: Chichester, UK, 2017; pp. 140–162. [CrossRef]
4. Miranda, K.; Millán, M.C.; Haros, C.M. Effect of Chia as Breadmaking Ingredient on Nutritional Quality, Mineral Availability, and Glycemic Index of Bread. *Foods* **2020**, *9*, 663. [CrossRef] [PubMed]
5. Orden de 31 de Enero de 1977 por la que se Establecen los Métodos Oficiales de Análisis de Aceites y Grasas, Cereales y Derivados, Productos Lácteos y Productos Derivados de la Uva. Boletín Oficial del Estado, 167, de 14 de Julio de 1977. Available online: https://www.boe.es/eli/es/o/1977/01/31/(1)/con (accessed on 10 May 2023).
6. *UNE-EN 14111:2022*; Derivados de Aceites y Grasas. Ésteres metílicos de ácidos grasos (FAME). Determinación del índice de yodo; Asociación Española de Normalización y Certificación: Madrid, Spain, 2022.
7. *UNE-EN ISO 3656:2011/A1:2017*; Aceites y Grasas de Origen Animal y Vegetal. Determinación de la absorbancia ultravioleta expresada como extinción UV específica; Asociación Española de Normalización y Certificación: Madrid, Spain, 2017.
8. Zhang, M.; Yang, X.; Zhao, H.T.; Dong, A.J.; Wang, J.; Liu, G.Y.; Zhang, H. A quick method for routine analysis of C18 trans fatty acids in non-hydrogenated edible vegetable oils by gas chromatography–mass spectrometry. *Food Control.* **2015**, *57*, 293–301. [CrossRef]
9. Härtig, C. Rapid identification of fatty acid methyl esters using a multidimensional gas chromatography–mass spectrometry database. *J. Chromatogr. A* **2008**, *1177*, 159–169. [CrossRef] [PubMed]
10. Eckhoff, S.R.; Singh, S.K.; Zehr, B.E.; Rausch, K.D.; Fox, E.J.; Mistry, A.K.; Keeling, P.L. A 100-g laboratory corn wet-milling procedure. *Cereal Chem.* **1996**, *73*, 54–57.
11. Real Decreto 308/1983, de 25 de Enero, por el que se Aprueba la Reglamentación Técnico-Sanitaria de Aceites Vegetales Comestibles. Boletín Oficial del Estado, 44, de 21 February 1983, Spain. Available online: https://www.boe.es/eli/es/rd/1983/01/25/308/con (accessed on 10 March 2023).

Proceeding Paper

Development of Extruded Snacks with Protein Hydrolysed from Jumbo Squid (*Dosidicus gigas*) by-Product and Cañihua (*Chenopodium pallidicaule Aellen*) †

Mateo Tapia *, Sebastián J. Marimón and Nicolás Salazar

Carrera de Ingeniería Industrial, Universidad de Lima, Av. Javier Prado Este 4600, Surco, Lima 15023, Peru; 20192310@aloe.ulima.edu.pe (S.J.M.); nsalazar@ulima.edu.pe (N.S.)

* Correspondence: author: 20191971@aloe.ulima.edu.pe

† Presented at the V International Conference la ValSe-Food and VIII Symposium Chia-Link, Valencia, Spain, 4–6 October 2023.

Abstract: The jumbo squid fishery in Peru is the second most important after the anchovy fishery. During its manufacturing process, only 50% to 60% of the total jumbo squid is used, thereby, the residues could be used to develop functional foods. Cañihua (*Chenopodium pallidicaule* Aellen) is an Andean pseudocereal from the highlands of Peru characterized by its high nutritional value. This work aimed to develop extruded snacks with protein hydrolyzed (PH) from jumbo squid by-product (JSBP) due to its high protein content, low price, and high availability. Four extruded snacks with corn flour (55%), rice flour (20% to 30%) and cañihua flour (15%) were enriched with PH from JSBP (4% to 10%) and developed using a twin-screw extruder. The extruded snacks were characterized by their physical properties (density, expansion ratio, water absorption index) and shelf life. The addition of PH from JSBP increased the protein content from 11.20% to 15.39%; ash content from 1.40% to 1.66% and fat content ranged from 1.10% to 1.18% compared to the control sample, the moisture content was from 4.46% to 5.81%. Also, the extruded snacks showed high phenolic concentration, 5633 µg GAE/g snack to 7315 µg GAE/g snack, high antioxidant activity, 698 mg trolox/g snack to 1274 mg trolox/g snack, high in vitro protein digestibility, 72.58% to 74.40%, and low acid index (0.095 mg KOH/g snack to 0.105 mg KOH/g snack) and peroxide index (0.00 meq O_2/kg snack to 0.063 meq O_2/kg snack), respectively. The snacks were accepted by the panel evaluators, complied with the Peruvian standard NTP-209.226 and microbiological requirements. Therefore, these snacks can be a healthier alternative product and satisfy market trends.

Keywords: Jumbo squid; cañihua; extruded snack; protein hydrolysate; antioxidants

Citation: Tapia, M.; Marimón, S.J.; Salazar, N. Development of Extruded Snacks with Protein Hydrolysed from Jumbo Squid (*Dosidicus gigas*) by-Product and Cañihua (*Chenopodium pallidicaule Aellen*). *Biol. Life Sci. Forum* **2023**, *25*, 4. https://doi.org/10.3390/blsf2023025004

Academic Editors: Claudia M. Haros, Loreto Muñoz and María Dolores Ortolá

Published: 28 September 2023

1. Introduction

Commercial snacks expanded by extrusion are foods with great acceptance by the consumer; however, these snacks are high in calories, low in protein content, minerals, dietary fiber, essential amino acids, and other bioactive compounds. The extrusion technology can be used to manufacture a wide range of food products with low processing costs, continuous production, high throughputs, and better product quality with optimal energy utilization [1]. Food extrusion is a process of several unit operations as thermal and mechanical treatment by which protein and/or starch-rich ingredients are fed into an extruder barrel through the feed hopper and the screw then conveys the food by compression, mixing, shearing, kneading and high temperature to produce a variety of products [2]. Several studies have shown that is possible to develop extruded snacks with pseudocereals such as quinoa to improve their antioxidant properties and with jumbo squid mantle, due to its high protein content, low price, and high availability [3–5]. Canihua, (*Chenopodium pallidicaule* Aellen), is a nutritious grain from the South American Andean highlands, cultivated in Puno, Peru, between 3200 and 4200 amsl. The cañihua cultivars present higher protein content (up to 20% of Dry Mass (DM)) and lipids content (up to 10% DM), crude fiber, and mineral contents compared to common cereals such as rice, corn, barley, and pseudocereals such as amaranth or quinoa [6]. Jumbo squid (*Dosidicus gigas*), also known as Humboldt squid, is one of the largest cephalopods and lives in the eastern Pacific Ocean. It represents an important economic fishery resource in Peru and other countries. Nevertheless, only the jumbo squid mantle is marketed, thereby, large amounts (up to 60% of whole weight) of squid off-products, such as skin, heads, fins, tentacles, and guts are generated and discarded, which are rich in many nutrients (proteins, lipids, minerals, biopolymers, etc.) [7]. There are no reports of the addition of protein hydrolyzed from jumbo squid (*Dosidicus gigas*) By-Product in extruded foods as far as we know. The objective of this research was to develop and characterize extruded snacks with protein hydrolyzed from JSBP and cañihua (*Chenopodium pallidicaule Aellen*).

2. Material and Methods

2.1. Raw Material

Jumbo squid By-Products (JSBP) (*Dosidicus gigas*) were collected from the Sechura Bay, Piura Department-Peru. Andean pseudocereal cañihua (*Chenopodium pallidicaule Aellen*), corn and rice flours were obtained from the local market in Lima city. JSBP were thawed, washed, and dehydrated by an infrared dryer (IRC D18, Irconfort, Sevilla, Spain) at 60 °C for 12 h in the Functional Foods Laboratory from the Universidad de Lima-Peru, and they were ground into the food shredder (Grindomix GM200, Restch, Haan, Germany) to obtain JSBP flour. The protein hydrolysed (PH) from JSBP flour was optimized and characterized using response surface methodology with a Box-Behnken design [8]. The parameters to optimize were temperature (40 °C to 60 °C), time (1 h to 2 h), and pH (5 to 8), the enzyme used was flavouryzme and 30 runs were carried out. The optimized parameters for the extraction of PH from JSBP were at 46° C for 1 h, 40 min and pH 6.5. All the samples were kept in aluminised bags at room temperature for later analysis.

2.2. Samples Preparation

Four extruded snacks were developed with different amounts of PH from JSBP and cañihua (Table 1). The ingredients of each sample were homogenized using a planetary mixer (FPSTSMPL1-053, Oster, Guangdong, China) and they were kept in aluminized bags at room temperature.

Table 1. Formulations for extruded snacks (ES) with protein hydrolyzed (PH) from jumbo squid By-products (JSBP) with cañihua (*Chenopodium pallidicaule* Aellen).

Samples	Corn Flour (%)	Rice Flour (%)	Cañihua Flour (%)	PHJSBP (%)
Control	55	30	15	-
ES1	55	26	15	4
ES2	55	24	15	6
ES3	55	22	15	8
ES4	55	20	15	10

2.3. Extrusion Process

Extrusion was performed in a twin-screw extruder scale pilot plant Model SLYE32-II (Saibainuo, China) at 10 hz of feeder speed and 400 rpm of screw speed. The extrusion process temperatures were 60, 90, 120 and 150 °C. The sample was fed to a mass flow of 200 g/min. The extruded products were stored at ambient conditions (25 °C, RH = 65%) in polyethylene bags until use. All the extrusion parameters were monitored during the process [4].

2.4. Physicochemical Characterization

2.4.1. Expansion Index (EI) and Bulk Density

The expansion index (EI) was measured by dividing the diameter of 15 extruded snacks by the diameter of the nozzle at the exit of the extruder (5 mm); the bulk density of the extrudates was determined using the seed (quinoa) displacement method [4]. The determinations were performed by triplicate.

2.4.2. Water Absorption Index (WAI)

A centrifuge tube was placed 1 g of extrudate with 50 mL of distilled water, left to stand for 30 min at room temperature and centrifuged for 15 min at 4000 rpm using a Centrifuge (HettichZentrifugen-Mikro, Germany). The WAI was the value of the weight of the gel obtained after removal of supernatant per unit weight of dry solids originals [4].

2.4.3. Proximal Composition

The moisture content of the extruded snacks was determined at 110 °C to a constant weight. The ash content was determined by ignition method (550 °C for 72 h). The fat content was determined with hexane for 9 h and the total Protein content was determined as % nitrogen × 6.25 using a Kjeldahl analyzer (UDK 139, VELP, Usmate Velate. Italy), by official methods.

2.4.4. Total Phenolic Content (TPC)

The total phenolic content (TPC) of the extruded snacks was determined by Folin–Ciocalteau method [9] at 760 nm using a spectrophotometer (UV 1280 Vis Spectrophotometer Shimadzu, Kyoto, Japan). The results were expressed as μg of gallic acid equivalent (GAE)/g snack. Analyzes were performed in triplicate and presented as mean values.

2.4.5. Antioxidant Activity

The antioxidant activity of samples was determined by the DPPH method [9] with some modifications at 517 nm by spectrometry, and ABTS free radical scavenging assay [9] at 734 nm (UV 1280 Vis Spectrophotometer Shimadzu, Kyoto, Japan). The results were expressed as μg Trolox/g snack. Analyzes were performed in triplicate and presented as mean values.

2.4.6. In Vitro Protein Digestibility (IVPD)

In vitro protein digestibility was measured by the method of Tinus et al. (2012) [10], with slight modification. The results will be expressed as:

$$\text{IVPD (Digestibility (\%))} = 65.66 + 18.10 \times (\text{pH 0 min} - \text{pH 10 min})$$

2.5. Determination of Amino Acid Profile

The amino acid profile was estimated according to Alaiz et al. [8] with slight modifications. Amino acid profile were determined by HPLC (ARC, Waters, Milford, CT, USA) with an inner diameter of 150 mm × 9 mm, column C18 reverse phase and tryptophan was quantified by HPLC after basic hydrolysis. All measures were performed in duplicate.

2.6. Oxidative Stability

The oxidative stability was determined using an 892 Professional Rancimat© (Metrohm, Herisau, Switzerland) [9]. To calculate the shelf life at 25 °C, induction periods at 80, 100, and 120 °C were determined. The determinations were performed in triplicate. Also, acid index (NTP 206.013) and peroxide index (NTP 201.016), were determined to evaluate the degree of hydrolysis and the oxidation state of the samples at room temperature.

2.7. Sensory Analysis

A panel of 50 individuals (university students between 18–26 years old) evaluated the sensory attributes for appearance, flavor, texture, and overall acceptability. The test was based on a 9-point hedonic scale (1 = dislike extremely and 9 = like extremely). The panelists received random samples served in plastic cup and where coded with 5 different digits [9].

2.8. Statistical Analysis

Results were expressed as mean ± standard deviation, all measurements were conducted in triplicate, except for the measurement of oxidative stability and the evaluation of gastrointestinal simulation, which were done in duplicate. Analysis of variance (ANOVA) was used to analyze the acquired data at a 95% significance level (Minitab Inc., USA).

3. Results and Discussions

According to Table 2, the expansion index (1.50 ± 0.05 cm/cm to 2.32 ± 0.05 cm/cm) and the diameter (7.50 ± 0.26 mm to 11.60 ± 0.26 mm) of the extruded snacks were significantly decreased when the content of PH from JSBP was increased (4% to 10%), while the density values (0.213 ± 0.034 g/cm^3 to 0.064 ± 0.008 g/cm^3) increased. These results were similar to those obtained by Roldan et al. [5]. Expansion indexes have an important role in the acceptability of the final product.

Table 2. Physical characteristics of extruded snacks (ES) with protein hydrolyzed (PH) from Jumbo Squid By-products (JSBP) and cañihua (*Chenopodium pallidicaule* Aellen).

Sample	Expansion Index (cm/cm)	Density (g/cm^3)	Diameter (mm)	WAI (g/g)
Control	2.49 ± 0.08	0.074 ± 0.005	12.47 ± 0.40	6.89 ± 0.23
ES1 (4% PHJSBP)	2.32 ± 0.05	0.064 ± 0.008	11.60 ± 0.26	6.40 ± 0.15
ES2 (6% PHJSBP)	1.83 ± 0.05	0.221 ± 0.047	9.17 ± 0.25	6.36 ± 0.28
ES3 (8% PHJSBP)	1.66 ± 0.05	0.226 ± 0.037	8.30 ± 0.26	6.23 ± 0.09
ES4 (10% PHJSBP)	1.50 ± 0.05	0.213 ± 0.034	7.50 ± 0.26	6.11 ± 0.15

Results are expressed as means ± SD (n = 3). WAI: Water Absorption Index.

The WAI of extruded snacks were significantly higher (6.11 ± 0.15 g/g to 6.40 ± 0.15 g/g) than those reported by Roldan et al. [4]. The increase of protein content may have caused the reduction in the capacity of starch to bind with water [11]. The physical characteristics of extruded snacks showed slight difference compared to control sample without PH from JSBP.

The proximate composition is presented in Table 3. By increasing the percentages of PH from JSBP, the protein content increased, being $15.39 \pm 0.12\%$ the highest in the ES4 sample compared to the control sample ($8.35 \pm 0.25\%$), as well as in comparison with other snacks, for example fish meal and corn grits (6.82% to 11.85%); fish powder, corn grits and rice grits (8.9% to 12.0%) [11]; and 13.34% for snacks enriched with jumbo squid mantle protein [5]. The lipid content ($1.10 \pm 0.07\%$ to $1.18 \pm 0.04\%$) and the ash content ($1.40 \pm 0.01\%$ to $1.66 \pm 0.04\%$) were higher than the control sample, meanwhile carbohydrates (80.35% to 77.57%) were lower. The moisture content is lower than $5.81 \pm 0.21\%$, this percentage is a quality index that indicates that the product will be less susceptible to the effect of microbes, therefore, the extruded snacks with PH from JSBP will have a long-life stability.

Table 3. Proximate composition of extruded snacks (ES) with protein hydrolyzed (PH) from Jumbo Squid By-products (JSBP) and cañihua (*Chenopodium pallidicaule* Aellen).

Sample	Moisture (%)	Lipids (%)	Protein (%)	Ash (%)	Carbohydrates (%)
CONTROL	6.04 ± 0.10	0.99 ± 0.01	8.35 ± 0.25	1.37 ± 0.01	83.25 ± 0.27
ES1 (4% PHJSBP)	5.81 ± 0.21	1.10 ± 0.07	11.20 ± 0.25	1.54 ± 0.01	80.35 ± 0.33
ES2 (6% PHJSBP)	5.35 ± 0.13	1.12 ± 0.01	12.27 ± 0.25	1.66 ± 0.04	79.60 ± 0.28
ES3 (8% PHJSBP)	4.85 ± 0.14	1.15 ± 0.05	14.14 ± 0.13	1.58 ± 0.06	78.28 ± 0.21
ES4 (10% PHJSBP)	4.46 ± 0.01	1.18 ± 0.04	15.39 ± 0.12	1.40 ± 0.01	77.57 ± 0.13

Results are expressed as means ± SD (n = 3).

The results of TPC, ABTS and DPPH are shown in Table 4. The extruded snacks showed a higher TPC (5633 ± 531 µg GAE/g snack to 7315 ± 521 µg GAE/g snack) compared with the control sample. The antioxidant activity by ABTS (3599 ± 112 µg Trolox/ g snack to 7626 ± 110 µg Trolox/ g snack) and by DPPH (5745 ± 500 µg Trolox/g snack to 12,739 ± 461 µg Trolox/g snack) were higher than the control sample.

Table 4. Total polyphenolic content (TPC) (µg GAE/g snack), ABTS (µg Trolox/g snack), DPPH (mg Trolox/g snack) and in vitro digestibility from extruded snacks (ES) with protein hydrolyzed (PH) from Jumbo Squid By-products (JSBP) and cañihua (*Chenopodium pallidicaule* Aellen).

Sample	TPC (µg GAE/g Snack)	ABTS (µg Trolox/g Snack)	DPPH (µg Trolox/g Snack)	In Vitro Digestibility (%)
CONTROL	4726 ± 542	2714 ± 114	5745 ± 500	74.76 ± 0.47
ES1 (4% PHJSBP)	5633 ± 531	3599 ± 112	6198 ± 480	72.58 ± 0.51
ES2 (6% PHJSBP)	6236 ± 521	4944 ± 190	9028 ± 481	73.40 ± 0.13
ES3 (8% PHJSBP)	6734 ± 531	6338 ± 112	10,519 ± 461	73.03 ± 0.39
ES4 (10% PHJSBP)	7315 ± 521	7626 ± 110	12,739 ± 461	74.40 ± 0.51

Results are expressed as means ± SD (n = 3).

According to Table 4, the extruded snacks presented high in vitro digestibility (72.58 ± 0.51% to 74.4 ± 0.51%) increased according to the amount of PH from JSBP. Extrusion has been reported to improve protein digestibility due to the protein denaturation and inactivation of antinutritional factors [7,11].

The amino acid composition of the extruded snacks showed a balanced profile, according to the FAO/WHO recommendations for healthy nutrition [8]. Taking into consideration the highest amino acids of ES4 (10% PHJSBP) were: aspartic acid + asparagine (83.88 ± 0.23 mg/g protein), glutamic acid + glutamine (147.67 ± 9.25 mg/g protein), glycine (52.26 ± 0.62 mg/g protein), alanine (64.32 ± 0.86 mg/g protein), valine (47.24 ± 0.46 mg/g protein), isoleucine (42.47 ± 0.45 mg/g protein), leucine (84.91 ± 0.51 mg/g protein), phenylalanine (41.72 ± 0.21 mg/g protein), and lysine (51.51 ± 0.24 mg/g protein).

The samples showed a low acid index (0.05 ± 0.00% of oleic acid) and peroxide index (0.00 ± 0.00 meq O_2/kg snack to 0.063 ± 0.01 meq O_2/kg snack), respectively. Also, the lifetime by the Rancimat© method showed an extrapolated shelf life at 25 °C of between 283 days to 630 days, which is higher than the control sample (248 days), this can be due to a high polyphenolic content and antioxidant activity from the addition of PH from JSBP.

The extruded snacks presented a light beige color with no residual smell of squid, similar results were reported by Roldan et al. [4,5]. In the sensory evaluation, consumers acceptability was considered above 5 (neither like nor dislike) in a hedonic scale from 1 to 9 [11]. Thus, an acceptability of 74% was obtained across all the samples, the characteristics of appearance, flavor, texture, and overall acceptability of the extruded snacks had an average value of 5.64, 4.98, 5.89, and 5.26, respectively. The sensory aspect would be one of the most important factors influencing eating behavior along with cost, safety, and accessibility.

4. Conclusions

The addition of PH from JSBP and cañihua improved the physical properties and increased the protein contented, phenolic concentration, antioxidant activity and in vitro protein digestibility of the extruded snacks compared with the control sample. Also, the results showed that the extruded snacks complied with the essential amino acids according to FAO / WHO. The snacks were accepted by the panel evaluators and complied with the Peruvian standard NTP-209.226, and they are much healthier than several commercially available snacks products. Therefore, the extruded snacks developed can be an alternative for the healthy food market and satisfy market trends.

Author Contributions: All authors have contributed equally to this manuscript. Conceptualization, M.T., S.J.M. and N.S.; Methodology, M.T., S.J.M. and N.S.; Investigation and Data analysis, M.T., S.J.M. and N.S.; Writing-original draft preparation, M.T., S.J.M. and N.S.; Writing-review and editing, M.T., S.J.M. and N.S. All authors have read and agreed to the published version of the manuscript.

Funding: This research received no external funding.

Institutional Review Board Statement: Not applicable.

Informed Consent Statement: Not applicable.

Data Availability Statement: Data is contained within the manuscript.

Acknowledgments: This work was supported by grant Ia ValSe-Food (119RT0567) and Laboratorio de Alimentos Funcionales of Carrera de Ingeniería Industrial, Universidad de Lima-Peru.

Conflicts of Interest: The authors declare no conflict of interest.

References

1. Grasso, S. Extruded snacks from industrial by-products: A review. *Trends Food Sci. Technol.* **2020**, *99*, 284–294. [CrossRef]
2. Abdullah, S.B.; Rosmin, M.J.; Reshma, K.; Sandeep, S.R.; Rama, C.P. Recent development, challenges, and prospects of extrusion technology. *Future Foods* **2021**, *3*, 100019. [CrossRef]
3. Valenzuela-Lagarda, J.L.; Pacheco-Aguilar, R.; Gutiérrez-Dorado, R.; Mendoza, J.L.; López-Valenzuela, J.Á.; Mazorra-Manzano, M.Á.; Muy-Rangel, M.D. Interaction of squid (*Dosidicus giga*) mantle protein with a mixtures of potato and corn starch in an extruded snack, as characterized by FTIR and DSC. *Molecules* **2021**, *26*, 2103. [CrossRef] [PubMed]
4. Roldán Acero, D.; Omote-Sibina, J.R.; Molleda Ordoñez, A.; Olivares Ponce, F. Desarrollo de barras nutritivas utilizando cereales, granos andinos y concentrado proteico de pota. *J. High Andean Res.* **2022**, *24*, 17–26. [CrossRef]
5. Roldán Acero, D.; Omote-Sibina, J.R.; Osorio-Lescano, C.M.; Molleda Ordoñez, A. Desarrollo de un producto extruido a base de cereales y concentrado de proteína de calamar gigante (*Dosidicus gigas*). *Intropica* **2021**, *16*, 34–42. [CrossRef]
6. Mérida-López, J.; Pérez, S.J.; Morales, R.; Purhagen, J.; Bergenståhl, B.; Rojas, C.C. Comparison of the Chemical Composition of Six Canihua (Chenopodium pallidicaule) Cultivars Associated with Growth Habits and after Dehulling. *Foods* **2023**, *12*, 1734. [CrossRef] [PubMed]
7. Ezquerra-Brauer, J.M.; Aubourg, S. Recent trends for the employment of jumbo squid (*Dosidicus gigas*) by products as a source of bioactive compounds with nutritional, functional and preservative applications: A review. *Int. J. Food Sci. Technol.* **2019**, *54*, 987–998. [CrossRef]
8. Chasquibol, N.; Gonzales, B.F.; Alarcón, R.; Sotelo, A.; Márquez-López, J.C.; Rodríguez-Martin, N.M.; del Carmen Millán-Linares, M.; Millán, F.; Pedroche, J. Optimisation and Characterisation of the Protein Hydrolysate of Scallops (Argopecten purpuratus) Visceral By-Products. *Foods* **2023**, *12*, 2003. [CrossRef] [PubMed]
9. Chasquibol, N.; Alarcón, R.; Gonzáles, B.F.; Sotelo, A.; Landoni, L.; Gallardo, G.; García, B.; Pérez-Camino, M.C. Design of Functional Powdered Beverages Containing Co-Microcapsules of Sacha Inchi P. *huayllabambana* Oil and Antioxidant Extracts of Camu Camu and Mango Skins. *Antioxidants* **2022**, *11*, 1420. [CrossRef] [PubMed]
10. Tinus, T.; Damour, M.; Van Riel, V.; Sopade, P.A. Particle size-starch–protein digestibility relationships in cowpea (Vigna unguiculata). *J. Food Eng.* **2012**, *113*, 254–264. [CrossRef]
11. Sutharut, J.; Khanitta, R. Development of direct expanded high protein snack products fortified with sacha inchi seed meal. *J. Microbiol. Biotechnol. Food Sci.* **2021**, *10*, 680–684.

biology and life sciences forum

Effect of Hydration on the Technological Properties of Andean Maize Native Whole-Grain Flour Dough and Bread [†]

Rita M. Miranda, Manuel O. Lobo and Norma C. Sammán *

Facultad de Ingeniería CIITeD UnJu-Conicet, Ítalo Palanca 10, San Salvador de Jujuy 4600, Argentina;
rmiranda@unju.edu.ar (R.M.M.); mlobo958@gmail.com (M.O.L.)

* Correspondence: normasamman@gmail.com

[†] Presented at the V International Conference la ValSe-Food and VIII Symposium Chia-Link, Valencia, Spain, 4–6 October 2023.

Abstract: Andean maize is produced in the Argentine Northwest and can be used in gluten-free bread formulations. Water has an important role in the technological properties of flour and consequently in gluten-free bread quality. Processing whole-grain flour is difficult but improves the final product's nutritional profile. The work aimed to evaluate the effect of different levels of water (100, 110, and 120% based on flour weight) on the flow properties of dough, textural properties and cooking of bread. For the formulation of the mold bread, Andean maize (*Capia* and *Bolita*) whole-grain flour was used. The flow properties of dough, weight losses and textural properties in breads were determined. The flow tests showed a drop in the consistency index with increasing dough hydration. In contrast, the flow index increased with respect to its initial value in both varieties of maize. The weight loss after baking tended to increase significantly from 4.60 to 5.8% with hydration increasing in *Capia* maize bread. However, this trend was not observed in *Bolita*. The hardness, springiness, gumminess and chewiness determined varied significantly only in *Bolita* maize bread due to the effects of hydration. More consistent dough resulted in harder, more elastic, rubbery and chewier bread.

Keywords: dough; gluten-free; Andean maize; technological properties

Citation: Miranda, R.M.; Lobo, M.O.; Sammán, N.C. Effect of Hydration on the Technological Properties of Andean Maize Native Whole-Grain Flour Dough and Bread. *Biol. Life Sci. Forum* **2023**, *25*, 5. https://doi.org/10.3390/blsf2023025005

Academic Editors: Claudia M. Haros, Loreto Muñoz and María Dolores Ortolá

Published: 28 September 2023

1. Introduction

Gluten-free breads have several textural and sensory defects because they lack a three-dimensional structure like gluten capable of retaining the incorporated air. The combination of various ingredients and additives (hydrocolloids, modified starches, protein isolates, among others) is the most widely used strategy to imitate the viscoelastic properties of gluten necessary to obtain gluten-free bread of acceptable quality [1].

Vidaurre-Ruiz et al. [2] studied four gluten-free bread formulations using two starches (corn and potato) and hydrocolloids (tara and xanthan gum), finding that depending on the starch source, due to their morphology and sticking characteristics, they can interact with hydrocolloids differently, affecting the rheological properties of dough and the baking and textural properties of baked goods. In addition, the chemical structure of the hydrocolloid also influences the mentioned properties. Subsequently, Vidaurre-Ruiz et al. [3] experimented with the replacement of potato starch with quinoa, kiwicha and tarwi flour in different proportions, showing that in multicomponent systems the physicochemical interactions between flour components affecting the rheological and textural properties of gluten-free dough, making the study of these systems more complex.

Various methods are used to evaluate the effect of varying the water content in gluten-free dough, such as farinographic, adjustment calculations based on the water absorption capacity of the ingredients, retro-extrusion, and flow tests. The gluten-free dough, made with different ingredients, can be considered a structured system and show different viscoelastic and flow properties depending on the water content and the composition of the raw material used [4]. There are few studies on the relationship between the baking properties of gluten-free whole-grain flours from Andean grains and the flow properties of their doughs for adjusting the water content.

The aim of this work was to evaluate the effect of different levels of water (100, 110, and 120% based on flour weight) on the flow properties of dough, textural properties and cooking of bread.

2. Materials and Methods

2.1. Materials

Andean maizes (Capia and Bolita varieties) grown in the Ocumazo-Humahuaca province of Jujuy, Argentina, was used. The grains were milled in a hammer mill (Polymix System PX-MFC) to obtain whole-grain flour with grain size <710 μm. The maize had 10.33% and 10.07% moisture content, respectively. Commercial xanthan gum (9.58% moisture content) was used.

2.2. Breadmaking Process

The mold bread was made with whole-grain native maize flour; initially, the flour was mixed for 1 min to achieve homogenization of the samples. For every 100 g of substituted flours, 100, 110 and 120 mL of water (30 °C), 1.5 g of previously activated commercial dry yeast, 1.5 g salt (NaCl), 3 g sugar, 3 g of milk powder, 10 g butter, 0.5% xanthan gum and 2% egg albumin were added and mixed to low speed in hand mixer for 5 min. The dough obtained was put in molds and placed in a fermentation chamber for 50 min at 30 °C and 80–90% relative humidity. The fermented dough was baked at 150 °C for 50 min. The bread's technological properties were evaluated 24 h after baking.

2.3. Technological Properties

2.3.1. Dough Rheological Properties

Dough samples for the rheological tests were prepared without adding yeast. The rheological measurements were conducted using a TA rheometer (DHR 10, TA Instrument, EE.UU). All measurements were carried out at 25 °C, using parallel plate geometry of 40 mm diameter and 1 mm gap. The dough sample was placed between the plates and the edges were carefully trimmed with a spatula. The flow experiments were conducted under steady-shear conditions with shear rates ranging from 0.01 to 50 L/s. The consistency index and flow behavior index or flow index values were obtained by applying the power law to the shear stress vs. shear rate curves. The measurements were made in duplicate.

2.3.2. Weight Loss (WL)

Weight loss or baking loss was computed as [initial loaf weight before baking—the loaf weight after 24 h baking × 100]/initial loaf weight before baking. The measurements were made in duplicate.

2.3.3. Textural Analysis

Crumb texture profile analysis (TPA) was performed using a texture analyzer (TAXT plus, Stable Micro System, Godalming, UK) equipped with a 5 kg load cell. An aluminum cylindrical probe with a P/35 (35.0 mm) was used for the bread; samples from the center of bread (thickness of 10 mm) were compressed to 50% of their original height. The test speed was 1 mm/s and the waiting time was 5 s. The measurements were made in quadruplicate.

2.4. Statistical Analysis

Data obtained were analyzed using the INFOSTAT software (Córdoba, Argentina, version 2017.1.2). The results were evaluated via analysis of variance with a significance level (α) 0.05, followed by the post hoc test with LSD Fisher.

3. Results and Discussion

3.1. Dough Properties

Figure 1A,B show the variation of the consistency and flow index due to the effect of the water content in the formulated gluten-free doughs.

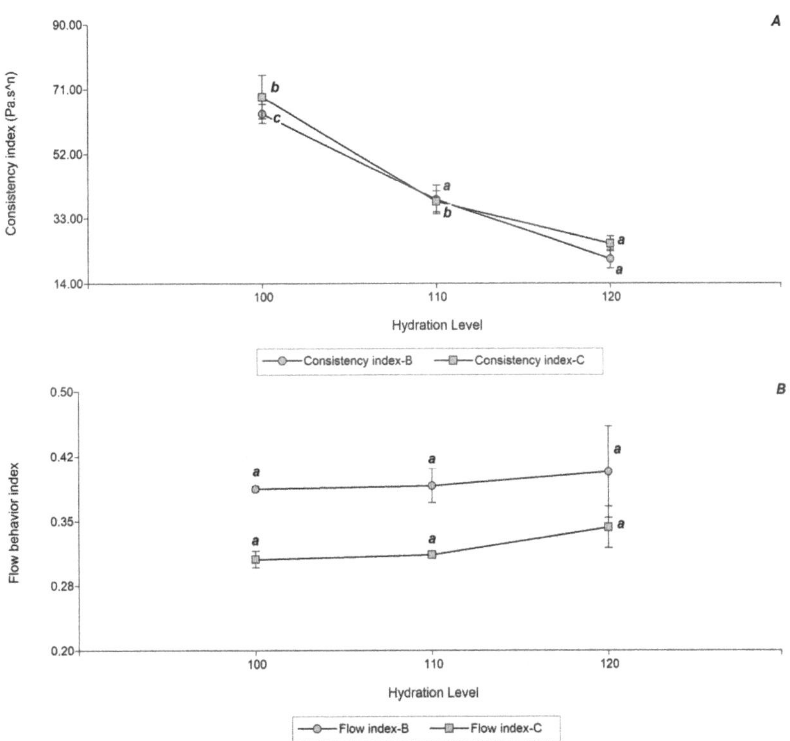

Figure 1. Diagram of values of (**A**) Consistency Index and (**B**) Flow Behavior Index of dough from Andean maize varieties *Bolita* (circle) and *Capia* (square), whit hydration levels 100, 110 and 120% in base of flour. Different letters indicate significant differences ($p < 0.05$) between hydration levels by maize.

The consistency of *Bolita* and *Capia* maize dough decreased significantly ($p < 0.05$) between 33.13 and 37.27%, respectively with increasing hydration. However, the values between maize did not vary significantly at the same level of hydration.

The flow behavior index did not show a significant difference with the hydration level in *Bolita* and *Capia* maize dough, although significant differences were observed between maize varieties ($p < 0.05$). The flow behavior index values obtained are within the range of values reported for other gluten-free dough formulations e indicate that the formulated dough with the flours of both maize varieties presented a pseudo plastic or shear thinning behavior [4]. However, in other formulations, we found an increment in the flow behavior index with increasing hydration that would be related with the greater mobility of the flour components. In this work, the high variability of this factor at the point of greatest hydration could be due to the greater ease of incorporation of air by the egg albumen, which would cause greater instability of the mass due to coalescence. Dermikersen et al. [5] indicated the existence of this unstable behavior, with phase separation, in control dough prepared only with rice flour (without gum or emulsifier) and in dough with pectin.

The studied maize varieties have morphological differences in their starch granules, with smaller sizes and greater size distribution in *Bolita* maize (unpublished results). In addition, their whole-grain flours have a compositional difference in lipid content (almost 20% higher in Capia) and fiber (almost 60% higher in Bolita). Despite this, they did not show differences in the flow behavior, possibly the native state of the whole-grain flours and the low stability provided by xanthan gum in the formulation dose.

3.2. Technological Characteristics of Bread

The weight losses (WL) by baking of the formulated breads are shown in Figure 2. In *Capia* maize WL increased from 4.60 to 5.8% in a statistically significant way ($p < 0.05$) with the level of hydration. In the *Bolita* maize bread, a trend similar to that of *Capia* maize at 100 and 110% hydration was observed, with a drop in WL to 120%, due to effect of water retention by the fiber. In some works, a direct relationship between bread quality and WL was observed; lower WL gave rise to loaves of lower volumes and defective textural characteristics [6]. In addition, the association of higher water contents with higher WL was highlighted, as observed in this work. The textural properties of the formulated bread can be shown in Figure 3. It was observed that increasing the hydration of the doughs had no significant effect on the textural properties of the *Capia* maize bread. In contrast, only a slight tendency to increase the hardness and elasticity of bread with 110% hydration was observed. In *Bolita* maize, a statistically significant decrease ($p < 0.05$) was observed in hardness, gumminess, chewiness, and elasticity at 100 and 110% hydration. No significant differences were found in cohesion or resilience in any sample. Higher values of elasticity, cohesion, and resilience in the bread are indicators of structural stability [7]. Encina-Zelada et al. [6] found that gluten-free bread with a low water content has greater hardness, which is consistent with the results of this work.

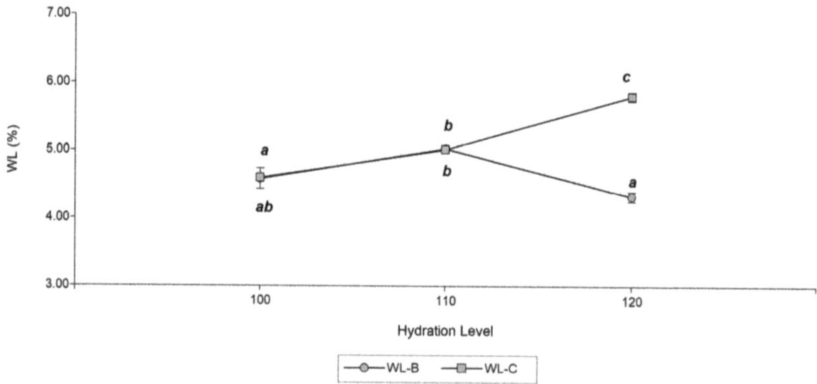

Figure 2. Diagram of values of weight loss (WL) of Andean maize varieties *Bolita* (circle) and *Capia* (square), whit hydration levels 100, 110 and 120% in base of flour. Different letters indicate significant differences ($p < 0.05$) between hydration levels of maize.

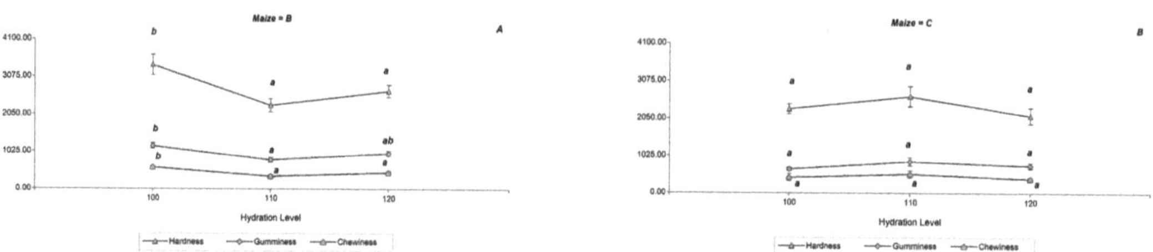

Figure 3. *Cont.*

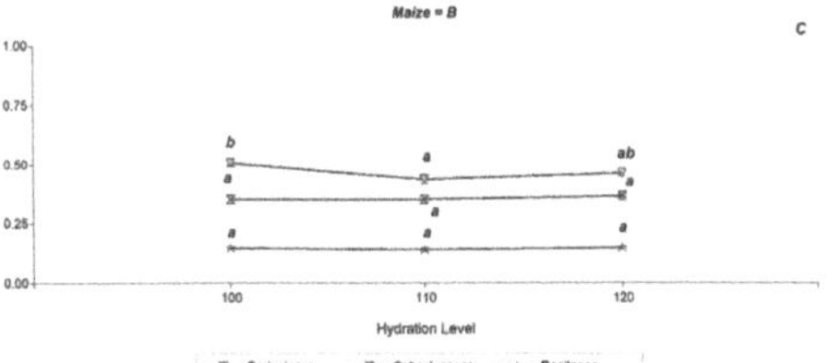
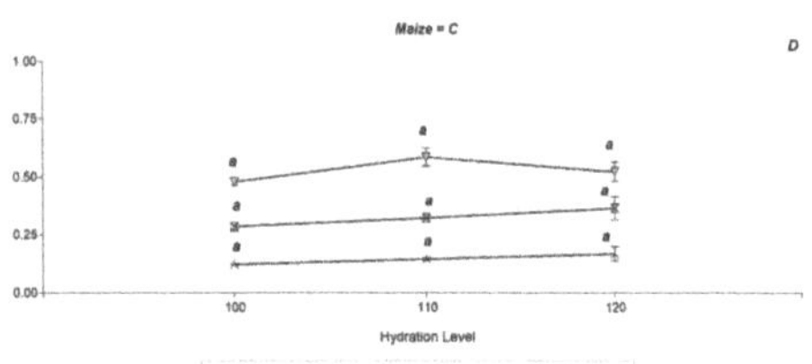

Figure 3. Diagram of values of Textural properties of Andean maize varieties *Bolita* (maize = B) and *Capia* (maize = C), whit hydration levels 100, 110 and 120% in base of flour. (**A,B**) Diagrams correspond to Hardness (triangle), Gumminess (rhombus) and Chewiness (pentagon); (**C,D**) diagram correspond to Springiness (inverted triangle), Cohesiveness (sandglass) and Resilience (star). Different letters indicate significant differences ($p < 0.05$) between hydration levels by maize.

That is, given the greater elasticity, the bread would be more structurally stable at 110% hydration in the case of *Capia* maize and at 100% hydration in the case of *Bolita* maize. However, these latter, due to their high hardness, would be less attractive to the consumer, and its dough is difficult to handle due to the high consistency. This agrees with the results of Santos et al. [8], who obtained higher acceptability in gluten-free chickpea bread with high hydration levels and low hardness.

4. Conclusions

The hydration level significantly influenced the consistency of the formulated dough independently of the variety of Andean maize used. On the other hand, the level of hydration significantly influenced the higher values of WL, hardness, elasticity, gumminess, and chewiness of *Bolita* maize breads at 100 and 110% water contents. In this case, the low structural stability of bread formulated with 120% water, flour in its native state, low gum concentration, and high fiber content of whole-grain flours, could be responsible for the drop in the mentioned parameters. The textural properties of the *Capia* maize dough did not show variations due to changes in water content, nor was a relationship found with flow properties.

Author Contributions: Conceptualization, M.O.L., R.M.M. and N.C S.; methodology, R.M.M., M.O.L. and N.C.S.; validation, R.M.M. and M.O.L.; formal analysis, R.M.M.; investigation, R.M.M. and N.C.S.; resources, M.O.L. and N.C.S.; data curation, R.M.M., M.O.L. and N.C.S.; writing—original draft preparation, R.M.M. and M.O.L.; writing—review and editing, M.O.L. and N.C.S.; visualization, M.O.L. and N.C.S.; supervision, M.O.L. and N.C.S.; project administration, N.C.S.; funding acquisition, M.O.L. All authors have read and agreed to the published version of the manuscript.

Funding: This research received no external funding.

Institutional Review Board Statement: Not applicable.

Informed Consent Statement: Not applicable.

Data Availability Statement: Not applicable.

Acknowledgments: This work was support by grant IaValSe-Food (119RT0567), CONICET and Secretaría de Ciencia y Técnica y Estudios Regionales, Universidad Nacional de Jujuy, Argentina.

Conflicts of Interest: The authors declare no conflict of interest.

References

1. Sciarini, L.S.; Steffolani, M.E.; Leon, A.E. El rol del gluten en la panificación y el desafío de prescindir de su aporte en la elaboración de pan. *Agriscientia* **2016**, *33*, 61–74. [CrossRef]
2. Vidaurre-Ruiz, J.; Matheus-Diaz, S.; Salas-Valerio, F.; Barraza-Jauregui, G.; Schoenlechner, R.; Repo-Carrasco-Valencia, R. Influence of tara gum and xanthan gum on rheological and textural properties of starch-based gluten-free dough and bread. *Eur. Food Res. Technol.* **2019**, *245*, 1347–1355. [CrossRef]

3. Vidaurre-Ruiz, J.; Salas-Valerio, F.; Schoenlechner, R.; Repo-Carrasco-Valencia, R. Rheological and textural properties of gluten-free doughs made from Andean grains. *Int. J. Food Sci. Technol.* **2021**, *56*, 468–479. [CrossRef]
4. Ronda, F.; Pérez-Quirce, S.; Villanueva, M. Rheological properties of gluten-free bread doughs: Relationship with bread quality. In *Advances in Food Rheology and Its Applications*, 1st ed.; Woodhead Publishing: Cambridge, UK, 2016; pp. 297–334. [CrossRef]
5. Demirkesen, I.; Mert, B.; Sumnu, G.; Sahin, S. Rheological properties of gluten-free bread formulations. *J. Food Eng.* **2010**, *96*, 295–303. [CrossRef]
6. Encina-Zelada, C.R.; Cadavez, V.; Monteiro, F.; Teixeira, J.A.; Gonzales-Barron, U. Combined effect of xanthan gum and water content on physicochemical and textural properties of gluten-free batter and bread. *Food Res. Int.* **2018**, *111*, 544–555. [CrossRef] [PubMed]
7. De La Hera, E.; Rosell, C.M.; Gomez, M. Effect of water content and flour particle size on gluten-free bread quality and digestibility. *Food Chem.* **2014**, *151*, 526–531. [CrossRef] [PubMed]
8. Santos, F.G.; Fratelli, C.; Muniz, D.G.; Capriles, V.D. The impact of dough hydration level on gluten-free bread quality: A case study with chickpea flour. *Int. J. Gastronomy Food Sci.* **2021**, *26*, 100434. [CrossRef]

Proceeding Paper

Comparison and Modeling of the Drying Kinetics of Moringa Leaves Using a Closed Facility in the Field and Using a Convective Tray Dryer [†]

María Dolores Ortolá [1], José Francisco García-Mares [2], Borja Mocholí [1], María Desamparados Soriano [3] and María Luisa Castelló [1,*]

[1] Food Engineering Research Institute (FoodUPV), Universitat Politècnica de València, 46022 Valencia, Spain; mdortola@tal.upv.es (M.D.O.); bormocpe@etsiamn.upv.es (B.M.)

[2] Department of Hydraulic Engineering and Environment, Universitat Politècnica de València, 46022 Valencia, Spain; fjgarcia@gmf.upv.es

[3] Department of Plant Production, Universitat Politècnica de València, Camino de Vera s/n, 46022 Valencia, Spain; asoriano@prv.upv.es

* Correspondence: mcasgo@upv.es

[†] Presented at the V International Conference la ValSe-Food and VIII Symposium Chia-Link, Valencia, Spain, 4–6 October 2023.

Abstract: Climate change requires a transition to crops that need less water and are more tolerant of high temperatures. In this regard, the adaptation of *Moringa oleifera* to the Spanish Mediterranean Basin offers the sustainable alternative production of a plant with a high nutritional value. It might also serve as a substitute crop with significant economic potential in less developed tropical regions in the world. Moreover, to provide the market with a homogeneous product, the stabilization of the leaves is necessary. The lack of control in the traditional form of moringa leaf drying (which only makes use of shade and air) does not guarantee a final homogeneous water content that can extend the crop's shelf life. Consequently, the purpose of this study was to develop a modular, affordable, and expandable dryer that enables the dehydration of leaves in the same field, lowering production costs in the process. The drying kinetics of leaves from crops of different ages (1 and 4 years) have been fitted to several mathematical models, using the "field dryer" and a semi-industrial tray dryer. In addition, their physicochemical properties were also compared. The outcomes demonstrate the viability of the field dryer's design. The drying kinetics of both dryers were more effectively adapted to the logarithmic model. Due to the variable air conditions in the field, the equilibrium moisture level attained was somewhat greater using the field dryer than using the tray dryer, but the product color and antioxidant and protein contents were similar. Finally, younger plants produced leaves with greater antioxidant capacities and a lower final water content.

Keywords: color; drying; moringa; leaves; modeling; protein

Citation: Ortolá, M.D.; García-Mares, J.F.; Mocholí, B.; Soriano, M.D.; Castelló, M.L. Comparison and Modeling of the Drying Kinetics of Moringa Leaves Using a Closed Facility in the Field and Using a Convective Tray Dryer. *Biol. Life Sci. Forum* **2023**, *25*, 6. https://doi.org/10.3390/blsf2023025006

Academic Editors: Claudia M. Haros and Loreto Muñoz

Published: 28 September 2023

1. Introduction

Moringa leaves contain many essential nutrients, e.g., vitamins, minerals, amino acids, beta-carotene, antioxidants, anti-inflammatory, anti-nutrients, and omega 3 and 6 fatty acids [1]. Drying these leaves is a common technique designed to extend their shelf life and also to guarantee homogeneous conditions for the introduction of dried moringa leaf powder in other food matrices, mainly in the bakery sector [2]. Furthermore, this powder can be used as a fortifier, in fish feed, or in livestock fodder [3].

The aim of this study was to model the drying kinetics of moringa leaves from 1- or 4-year-old crops using a field dryer or a tray dryer. In addition, the color, protein content, and antioxidant capacity were recorded.

2. Materials and Methods

2.1. Raw Materials

Moringa oleifera leaves collected during the months of October and November 2021 from specimens planted in 2016 and 2021, in the experimental plot of the Universitat Politècnica de València, Spain, were used for this research.

2.2. Drying Conditions and Equipment

In order to stabilize and prolong the shelf life of the moringa leaves, they were subjected to two types of dehydration for 30 h: in a tray dryer with a controlled relative humidity (35%) and temperature (40 °C) (Pol-Eko Aparatura, Clk 750 Top+), using 0.67 kg of moringa leaves/m^2 tray; and in a field dryer designed in a modular format. Specifically, it comprised two overlapping plastic pallet boxes perforated at the base, a lid, a retractable tube for air removal, an air extraction fan, an air heater (Voltomat Heating Industrial Heater), a temperature controller thermostat (HiLetgo XH-W3001 220V Digital LED Temperature Controller Thermostat), a temperature probe, a plug-in timer, a plastic tarpaulin, microperforated baskets, and dataloggers for recording the temperature and relative humidity values.

2.3. Drying Modeling

The drying kinetics were adjusted to the different models selected because they had been used in other vegetable products [4], in which the driving force * $\left(Y = \dfrac{\left(X_w^e - X_w^t\right)}{\left(X_w^e - X_w^0\right)} \right)$ was related to the time (t) by means of different coefficients. For this purpose, the variations in the weight and moisture of the leaves were obtained at different drying times (0, 1, 2, 3, 5, 7, 24, 26, 28, and 30 h).

* X_w is the amount of water in dry basis, e is equilibrium, 0 denotes the initial conditions, and t is the time.

2.4. Analytical Determinations

The moisture content of moringa leaves was determined via a gravimetric method, and their protein content was obtained via Kjeldahl methodology. Furthermore, the antioxidant capacity was registered via DPPH, and the CIEL*a*b* coordinates were recorded using a spectrophotometer ("Konica Minolta" Inc. Model CM—3600d, Tokyo, Japan) with a reference illuminant D65 and a 10° observer.

2.5. Statistical Analysis

Statgraphics Centurion software was used for the statistical analysis of the results. ANOVA (analysis of variance) was performed using the LSD (least significant difference) test at a significance level of 95% (*p*-value $\leq$ 0.05).

3. Results and Discussion

Figure 1 shows the experimental values of the driving forces with respect to the time in the moringa leaf drying process using both dryers and with moringa crops of differing ages, along with the fitting of the logarithmic model (Y = a·e^{-kt} + c), which exhibited the highest R^2 ($\geq$99% for the tray dryer and $\geq$96% for the field dryer). Focusing on its parameters (Table 1), we can see that the value of k (the drying constant) was higher for leaves dried in the field dryer than in the tray dryer, without any effect from the plantation age. As this parameter is associated with the effect of external conditions on drying, the greater variability in these conditions is evident in the field dryer. As was expected, the drying kinetic in the field dryer was slightly slower than in the tray dryer, as the temperature conditions were not as homogeneous at night. Thus, the moisture content in the moringa leaves after 30 h in the field dryer was ≈48%, whereas in the tray dryer, it was ≈40%. Furthermore, the younger the crop, the lower the water content in the leaves at the end of the drying process.

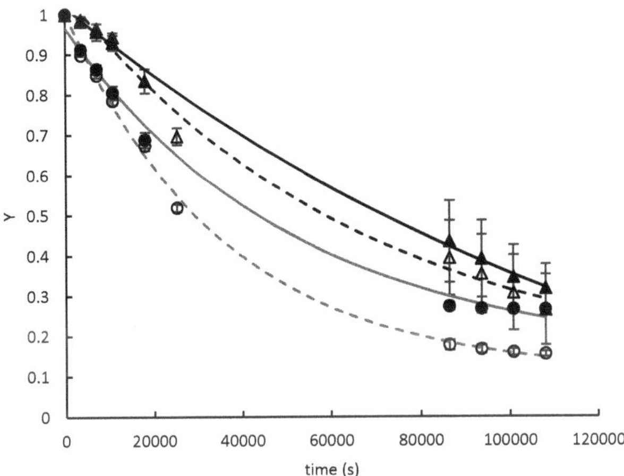

Figure 1. Evolution of the driving force (Y) versus drying time (s) of moringa leaves: the logarithmic model (continuous lines, 4-year-old moringa; dashed lines, 1-year-old moringa; black, field dryer; grey, tray dryer) and experimental points (triangles, field dryer; circles, tray dryer; filled symbols, 4-year-old moringa; empty symbols, 1-year-old moringa). Y, the driving force.

Table 1. The parameters of the logarithmic model for the drying kinetics of moringa leaves.

Plantation Age	Type of Dryer	Logarithmic Model Parameters	
1	trays		2.8×10^{-5}
4	trays		1.9×10^{-5}
1	field	k	1.47×10^{-5}
4	field		6.90×10^{-6}
1	trays		0.90
4	trays		0.82
1	field	a	0.96
4	field		1.33
1	trays		0.1026
4	trays		0.1399
1	field	c	0.094
4	field		−0.311

The protein content of the fresh moringa leaves was 10.8 ± 0.5%, with no significant differences as a result of the plantation age. After dehydration, a protein content of about 40% was reached, and neither the type of the dryer nor the age of the plantation was found to have a significant effect.

The % of DPPH inhibition was significantly higher in the dried moringa leaves from 1-year-old crops (≈50%) than from 4-year-old crops, and the type of dryer used had no significant effect.

Figure 2 shows the values of the CIELa*b* coordinates of the dried moringa leaves used in this study. Drying, regardless of the method and the crop age, increased the luminosity of the samples. The leaf color was placed in the second quadrant of the chromatic diagram, which corresponds to the area of yellows and greens. The fresh leaves exhibited a lower color purity, as they were located more in the center of the diagram. The difference between the field-dried and tray-dried leaves was less marked, with the latter being more greenish in tone.

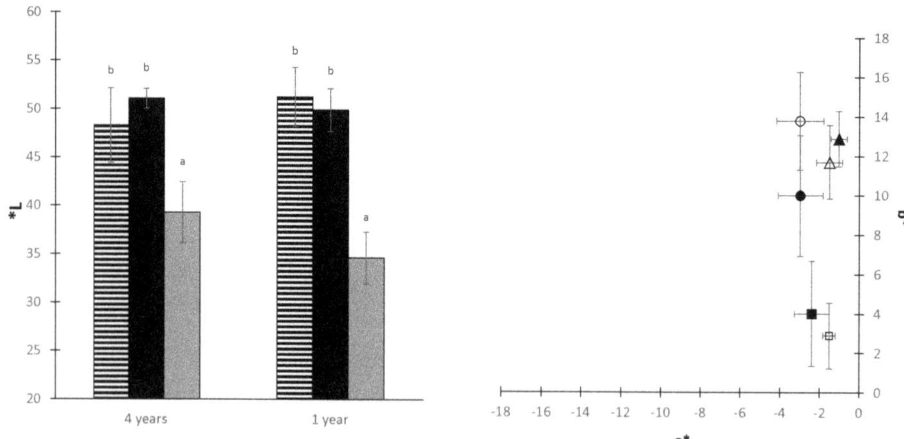

Figure 2. The luminosity (L*) (bars with horizontal lines, tray dryer; black bars, field dryer; grey bars, fresh leaves) and the location in the chromatic diagram (b* vs. a*) of the moringa dried leaves in this study (square, fresh leaves; triangles, field dryer; circles, tray dryer; filled symbols, 4-year-old moringa; empty symbols, 1-year-old moringa). Identical letters indicate homogeneous groups obtained in the ANOVA with n.s: 95%.

4. Conclusions

A modular and economical dryer placed in the same field in which moringa is produced could be a feasible way to enhance a circular economy, although the equilibrium humidity reached in the moringa leaves was slightly higher than that obtained using the tray dryer because the air conditions were less homogeneous throughout the process. A logarithmic model is useful for the purposes of estimating the proper drying time. The dehydrated moringa leaves registered a high antioxidant capacity, especially the leaves from the 1-year-old plantation, with the type of dryer having no effect. Due to the high protein content of dried moringa leaves ($\approx$40%), they could be incorporated into different food matrices to improve the nutritional level. Finally, dehydration increased the luminosity and color purity of moringa leaves.

Author Contributions: Conceptualization, M.D.O., J.F.G.-M., M.D.S. and M.L.C.; methodology, J.F.G.-M. and B.M.; validation, M.D.O. and M.L.C.; formal analysis, B.M., M.D.O. and M.L.C.; investigation, B.M. and J.F.G.-M.; resources B.M., M.D.O., M.L.C.; data curation, M.D.O. and M.L.C.; writing—original draft preparation, B.M..; writing—review and editing, M.D.O. and M.L.C.; supervision, M.D.O., M.L.C., J.F.G.-M. and M.D.S. funding acquisition, M.D.O., M.D.S. and M.L.C. All authors have read and agreed to the published version of the manuscript.

Funding: This research was funded by the project "Mejora de la producción y calidad de hojas de moringa en Paraguay para contribuir al aporte nutricional de grupos desfavorecidos (MORNUPAY)" (Ref. AD2115—Centre for Development Cooperation (CCD)—Universitat Politècnica de València, Spain). The APC was funded by la ValSe-Food grant (119RT0567).

Institutional Review Board Statement: Not applicable.

Informed Consent Statement: Not applicable.

Data Availability Statement: Data available on request from the authors.

Acknowledgments: The authors thank the support provided by the Universitat Politècnica de València and la ValSe-Food.

Conflicts of Interest: The authors declare no conflict of interest.

References

1. Essa, M.M.; Subash, S.; Parvathy, S.; Meera, A.; Guillemin, G.J.; Memon, M.A.; Manivasagam, T. Brain health benefits of Moringa oleifera. *Food Brain Health* **2014**, *2*, 113–118.
2. El-Gammal, R.; Ghoneim, G.; ElShehawy, S. Effect of Moringa Leaves Powder (Moringa oleifera) on Some Chemical and Physical Properties of Pan Bread. *J. Food Dairy Sci.* **2016**, *7*, 307–314. [CrossRef]
3. Abd El-Hack, M.E.; Alagawany, M.; Elrys, A.S.; Desoky, E.S.M.; Tolba, H.M.N.; Elnahal, A.S.M.; Elnesr, S.S.; Swelum, A.A. Effect of forage *Moringa oleifera* L. (*Moringa*) on animal health and nutrition and its beneficial applications in soil, plants and water purification. *Agriculture* **2018**, *8*, 145. [CrossRef]
4. Onwude, D.I.; Hashim, N.; Janius, R.B.; Nawi, N.M.; Abdan, K. Modeling the Thin-Layer Drying of Fruits and Vegetables: A Review. *Compr. Rev. Food Sci. Food Saf.* **2016**, *15*, 599–618. [CrossRef]

biology and life sciences forum

Influence of Substitution of Wheat and Broad Bean Flour for Hydrolyzed Quinoa Flour on Cookie Properties [†]

Ileana de los A. Gremasqui [1], María A. Giménez [1], Manuel O. Lobo [1], Loreto Muñoz [2], María C. Zuñiga [3] and Norma C. Sammán [1,*]

[1] CIITED Centro Interdisciplinario de Investigaciones en Tecnología y Desarrollo Social del NOA—CONICET, Facultad de Ingeniería, Universidad Nacional de Jujuy, Italo Palanca 10, Jujuy 4600, Argentina; ileanagremasqui96@gmail.com (I.d.l.A.G.); malejandragimenez@gmail.com (M.A.G.); mlobo958@gmail.com (M.O.L.)

[2] Facultad de Ingeniería, Universidad Central de Chile, Santiago 8330601, Chile; loreto.munoz@ucentral.cl

[3] Departamento de Química Inorgánica y Analítica, Facultad de Ciencias Químicas y Farmacéuticas, Univesidad de Chile, Sergio Livingstone, 1007, Independencia, Santiago 8380492, Chile; mczuniga@ciq.uchile.cl

[*] Correspondence: author: normasamman@gmail.com

[†] Presented at the V International Conference la ValSe-Food and VIII Symposium Chia-Link, Valencia, Spain, 4–6 October 2023.

Abstract: Quinoa (*Chenopodium quinoa*) is an important pseudocereal for its high nutritional value, versatility in cooking, gluten-free nature, and potential contribution to food security and sustainable agriculture. The aim of this work was to evaluate the effect of different levels of substitution (10, 20, and 30%) of hydrolyzed quinoa flour (HQF) on the nutritional, physical, and antioxidant characteristics and protein digestibility of cookies elaborated with wheat and broad bean flours. Cookies without HQF were the control (C0). The addition of HQF increased the protein content by between 12 and 68% compared to C0. The increase in HQF improved the cookies' quality according to the spread ratio. Adding HQF resulted in more compact cookies, decreasing their specific volume (1.30 to 1.15 cm^3/g) and increasing their hardness (2791 to 6515 g). The total polyphenols increased by 2 to 3 times, and the antioxidant activity increased by more than three times with a 30% addition of HQF with respect to C0. The oxygen radical absorbance capacity with fluoresceine (ORAC-FL) index (stoichiometry or amount of antioxidants) revealed that up to a 20% and 30% addition of HQF increased the antioxidant compounds by up to ~1.5 times. On the other hand, the antioxidant reactivity, according to the oxygen radical absorbance capacity with pyrogallol red (ORAC-PGR) index, increased by 2.4 times with a 30% addition of HQF. Finally, the cookies' digestibility improved with a 10% addition of HQF. Therefore, HQF represents a viable option in the development of cookies with highly reactive antioxidant compounds that are nutritionally improved. This application could be extended to other baked products. However, a 30% addition of HQF affects its textural properties and decreases its digestibility.

Keywords: *Chenopodium quinoa*; antioxidant; cookies; flour; hydrolyzed flour; properties; protein

Citation: Gremasqui, I.d.l.A.; Giménez, M.A.; Lobo, M.O.; Muñoz, L.; Zuñiga, M.C.; Sammán, N.C. Influence of Substitution of Wheat and Broad Bean Flour for Hydrolyzed Quinoa Flour on Cookie Properties. *Biol. Life Sci. Forum* **2023**, 25, 7. https://doi.org/10.3390/blsf2023025007

Academic Editors: Claudia M. Haros and María Dolores Ortolá

Published: 28 September 2023

1. Introduction

In recent years, the interest of the food industry in the development of functional food products and the attraction of consumers to these have increased. Protein-enriched bakery products, especially cookies, are widely accepted due to their adequate physicochemical, rheological, textural, antioxidant, and sensory properties and long shelf life [1]. Furthermore, a higher protein intake may provide benefits to people on special diets, such as the elderly and athletes.

Quinoa (*Chenopodium quinoa*) is an important pseudocereal for its high nutritional value, versatility in cooking, gluten-free nature, and potential contribution to food security and sustainable agriculture. There are several investigations on the effect of protein hydrolysates in bakery products [2]. However, studies on the incorporation of hydrolyzed quinoa to enrich these products is scarce. In our previous works, we studied the nutritional and functional properties of a hydrolyzed protein product from quinoa flour, which could be useful for the development of enriched baked products [3]. Therefore, the aim of this work was to evaluate the effect of different addition levels (10, 20, and 30%) of hydrolyzed quinoa flour (HQF), on the nutritional physical, antioxidant, and protein digestibility characteristics of cookies elaborated with wheat and broad bean flours.

2. Materials and Methods

2.1. Materials

Broad bean flour (BF) (*Vicia faba* L.) was provided by small producers from the Quebrada de Humahuaca, Jujuy, Argentina. Hydrolyzed quinoa flour (HQF), obtained in previous work (54.7% protein, 14.11% ash and 23.18% dietary fiber), was used [3]. Refined wheat flour (WF), baking powder, sugar, margarine, bitter cocoa, and vanilla essence was purchased from a local market.

2.2. Protein Cookies: Formulation

Protein cookies were formulated with a mixture of wheat flour (WF) and broad bean flour (BF) and the addition of different levels (10, 20, and 30%) of HQF (CQ10, CQ20, and CQ30). A control cookie (C0) was formulated with a mixture of WF/BF in a ratio of 70:30, respectively. Through preliminary tests, the amount of baking powder, sugar, margarine, bitter cocoa, and water to be incorporated was determined.

Cookies' preparation: The process began by creaming the margarine, sugar, and vanilla essence. When this step was completed, the flour mixture, baking powder, bitter cocoa, and water were added according to the ratios shown in Table 1. The dough formed was stretched and cut circularly with an approximate thickness of 8 mm. The dough circles were baked at 150 ± 2 °C for 30 min. Finally, the cookies were cooled at room temperature and then analyzed.

Table 1. Formulation of cookies (g/100 g).

Ingredients	C0 (g)	CQ10 (g)	CQ20 (g)	CQ30 (g)
WF	70	63	56	49
BF	30	27	24	21
HQF	0	10	20	30
Sugar	30	30	30	30
Margarine	30	30	30	30
Baking powder	3.2	3.2	3.2	3.2
Bitter cocoa	1	1	1	1
Water	20	18	18	22

WF: refined wheat flour; BF: broad bean flour; HQF: hydrolyzed quinoa flour; C0: control cookie; CQ10, CQ20, and CQ30: protein cookies elaborated with addition of 10, 20, and 30% of HQF.

2.3. Chemical Composition

AOAC (2017) methods were used to determine the moisture, protein, lipids, ashes, and total dietary fiber (TDF) of baked cookies. Digestible carbohydrate (DHC) content was calculated by difference.

2.4. Physical Properties

2.4.1. Diameter, Thickness, and Spread Ratio

The diameter and thickness of baked cookies were measured using a digital vernier caliper. These parameters were measured 4 times, rotating the cookie 90° after each measurement. An average of six baked cookies was determined. The spread ratio was determined from the ratio between the average values of the diameter and thickness of the cookies using the following equation:

$$\text{Spread ratio} = \text{Diameter}/\text{Thickness}$$

2.4.2. Specific Volume (SV)

The volume was determined according to Encina-Zelanda et al. [4] using a modified standard displacement method with quinoa seeds. The specific volume was calculated using the following equation:

$$\text{SV (cm}^3/\text{g)} = \text{Cookie volume}/\text{Cookie weight}$$

2.4.3. Hardness

Hardness was determined using a Texture Analyzer (TA-XT Plus, Stable Micro Systems Godalming, UK). The bending test was performed using a platform and 3-point bending ring (HDP/3 PB) with a 5 kg load cell. The hardness, peak of maximum force, was measured in six baked cookies for each formulation. The maximum force to break the cookies is reported as the hardness in N.

2.5. Antioxidant Properties

Antioxidant compounds were extracted using ultrasound-assisted extraction (UAE). Total phenolic content (TPC), 2,2-diphenyl-1-picrylhydrazyl (DPPH) free radical-scavenging assay, ORAC-FL (stoichiometry or amount of antioxidants), and ORAC-PGR (antioxidant reactivity) indexes were determined according to Zuñiga-López et al. [5].

2.6. In Vitro Protein Digestibility (IVPD)

The IVPD was determined using the standardized static in vitro digestion protocol, suggested by international consensus within the framework of INFOGEST COST Action [6].

2.7. Statistical Analysis

The data were statistically treated by analysis of variance, while the means were compared through the LSD Fisher's test at a significance level of 0.05 using, in both cases, the statistical software INFOSTAT—Version 2017 p (UNC, Cordoba, Argentina). All experiments were performed at least in triplicate, and mean values $\pm$ standard deviation were reported.

3. Results and Discussion

3.1. Chemical composition

The proximal composition on a wet basis (Table 2) shows that the protein content of the cookies increased from 9.66 to 16.23 g/100 g sample. This increase was significant and due to the 10% substitution of HQF. The ash content and total dietary fiber of the cookies increased with the addition of HQF. This could be due to the higher ash and fiber contents of HQF [3], so it could serve as an alternative source of dietary fiber in bakery products.

Table 2. Chemical composition of cookies (g/100 g sample).

Parameters *	C0	CQ10	CQ20	CQ30
Moisture	12.84 ± 0.17 [a]	11.45 ± 0.48 [b]	$9.49 \pm 0,21$ [c]	7.18 ± 0.65 [d]
Protein	9.66 ± 0.17 [d]	10.86 ± 0.07 [c]	$14.97 \pm 0,47$ [b]	16.23 ± 0.07 [a]
Lipids	12.46 ± 0.53 [a]	12.07 ± 0.07 [a]	$11.95 \pm 0,14$ [a]	11.83 ± 0.13 [a]
Ash	1.33 ± 0.24 [d]	2.39 ± 0.06 [c]	$3.09 \pm 0,08$ [b]	3.55 ± 0.19 [a]
TDF	7.12 ± 0.13 [d]	9.28 ± 0.24 [c]	$10.72 \pm 0,34$ [b]	12.15 ± 0.14 [a]
DHC	56.95	53.95	49.78	49.06

* Means $\pm$ standard deviations (n = 3). Values in each row followed by different superscript letters are significantly different ($p < 0.05$). TDF: total dietary fiber; DHC: digestible carbohydrates calculated by difference. C0: control cookie; CQ10, CQ20, and CQ30: protein cookies elaborated with addition levels of 10, 20, and 30% of hydrolyzed quinoa flour (HQF).

3.2. Physical Properties

Table 3 shows the physical properties of the cookies. Compared to C0, the diameter of the cookies increased due to the 20% substitution of HQF, while the thickness decreased significantly. Therefore, the spread ratio of the cookies increased significantly due to the 20% substitution of HQF. The decrease in total gluten content and the increase in protein content as HQF was added could be responsible for the increase in spread ratio. In this system, the water could interact with the proteins, peptides, and amino acids of the hydrolyzed flour and, consequently, be less available to interact with the gluten. The spread ratio is an important quality index in cookies, as those with higher spread ratio values are more desirable. Therefore, CQ30 would be considered the best quality with respect to this property.

Table 3. Physics properties of cookies.

Parameters *	C0	CQ10	CQ20	CQ30
Diameter (mm)	54.25 ± 0.81 [a]	55.06 ± 1.03 [a]	57.32 ± 1.51 [b]	57.73 ± 1.30 [b]
Thickness (mm)	11.84 ± 0.50 [b]	11.01 ± 0.78 [ab]	10.24 ± 0.20 [a]	10.08 ± 0.58 [a]
Spread Ratio	4.59 ± 0.23 [ab]	5.02 ± 0.40 [b]	5.60 ± 0.16 [c]	5.74 ± 0.20 [c]
SV (cm^3/g)	1.30 ± 0.02 [c]	1.20 ± 0.06 [b]	1.15 ± 0.12 [a]	1.15 ± 0.18 [a]
Hardness (g)	2791 ± 497 [a]	3056 ± 449 [b]	3788 ± 221 [c]	6515 ± 304 [d]

* Means ± standard deviations (n = 3). Values in each row followed by different superscript letters are significantly different ($p < 0.05$). SV: specific volume. C0: control cookie; CQ10, CQ20, and CQ30: protein cookies elaborated with addition levels of 10, 20, and 30% of hydrolyzed quinoa flour (HQF).

The specific volume (SV) decreased with the increasing HQF levels, indicating a more compact structure. This behavior could be explained by a strong interaction between free water, limited in the dough, and the proteins and fiber of the HQF. On the other hand, the decrease in SV could also be attributed to the dilution of the gluten content due to the addition of HQF. Furthermore, the fiber content provided by the HQF would weaken the dough, decreasing the SV [7].

In general, the cookies' hardness increased after the 10% addition of HQF. The increase in hardness could be related to the more compact structure due to the increase in protein and fiber content [8]. When comparing the hardness values with those of a commercial cookie (CC), which contained whey concentrate in its formulation (hardness: 9795.55 g), it was found that they were significantly lower.

3.3. Antioxidant Properties

Figure 1a,b show that the gradual addition of HQF significantly increases the TPC and antioxidant activity measured via the DPPH radical scavenging of the cookies compared to G0. On the other hand, the stoichiometry, or amount of antioxidants measured through the ORAC-FL index, increased by 1.5 times with the 20 and 30% substitutions HQF with respect to C0. Furthermore, the antioxidant reactivity against radicals determined with the ORAC-PGR index increased by 2.5 times with the 30% addition of HQF (Figure 1c).

The substitution with HQF caused an effective supplementation of phenolic compounds in cookies. According to Batista et al. [9], there is a high correlation between DPPH radical scavenging and total phenolic content. The increase in antioxidant activity would help to reduce the effect of free radicals and peroxides and would increase the potency of anti-oxidative enzymes in the body [10]. In addition, the Maillard reactions produced during baking led to an increase in molecules that reduce free radicals. These molecules can react with Folin Ciocalteu reagent, contributing to the overall antioxidant activity [11]. Espinosa-Páez et al. [12] reported that the bitter cocoa used as an ingredient in the formulation of cookies contains phenols and melanoidins that, when exposed to temperatures between 130 and 150 °C, increase the content of bound polyphenols and the antioxidant activity. Therefore, processing can increase the availability of antioxidant compounds and their antioxidant activity.

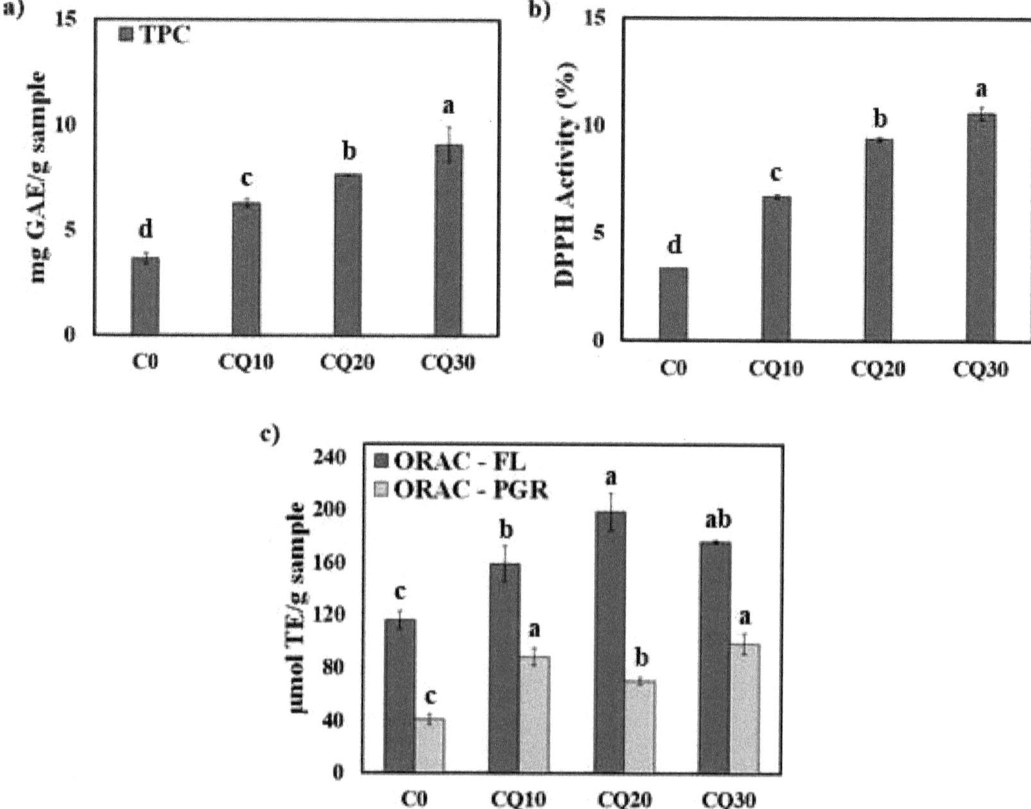

Figure 1. (**a**) Total phenolic content (TPC) of cookies. (**b**) Antioxidant activity of cookies measured by 2,2-diphenyl-1-picrylhydrazyl (DPPH) free radical-scavenging assay. (**c**) Antioxidant capacity measured by the oxygen radical absorbance capacity—Fluoresceine (ORAC-FL) and reactivity measured by the oxygen radical absorbance capacity—Pyrogallol red (ORAC-PGR) of cookies. Result are expressed as average ± standard deviation (n = 3). Bars with different lower-case letters are significantly different ($p < 0.05$). C0: control cookie; CQ10, CQ20, and CQ30: protein cookies elaborated with addition levels 10, 20, and 30% of hydrolyzed quinoa flour (HQF).

3.4. In Vitro Protein Digestibility (IVPD)

Figure 2 shows the IVPD of cookies as a function of time. Regarding G0, the IVPD increased significantly by 26% with the 10% addition of HQF. However, the addition of higher levels of HQF caused a decrease in IVPD, so no significant differences were observed between CQ20, CQ30, and G0. Bas and El [13] reported that the decrease in the IVPD is due to the Maillard reaction that occurred during baking and the methodology used to determine this property. Therefore, it is recommended to study other properties, such as the degree of denaturation and the microstructure of the food, which are factors that can affect protein digestibility.

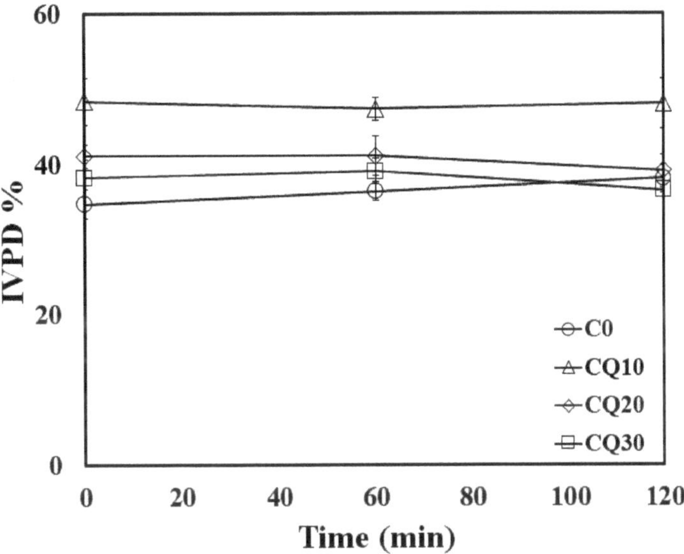

Figure 2. In vitro protein digestibility (IVPD) of C0 (o), CQ10 (Δ), CQ20 (◊), and CQ30 (□) as a function of time. Result are expressed as average ± standard deviation (number of repetition, n = 3). C0: control cookie; CQ10, CQ20, and CQ30: protein cookies elaborated with 10, 20, and 30% addition levels of hydrolyzed quinoa flour (HQF).

4. Conclusions

The results showed that the substitution with HQF notably increases the protein and fiber content of the cookies. The 30% addition of HQF increases the hardness of the cookies; however, this hardness value is lower than those of some commercial cookies. Furthermore, the addition of HQF increases the spreading ratio of the cookies, improving the cookies' quality. Cookies could be an alternative to functional food, and their consumption could be beneficial to people's health. On the other hand, the protein digestibility of cookies decreases with high levels of HQF. Therefore, the use of HQF represents a viable and advantageous option in the development of nutritionally enhanced bakery products.

Author Contributions: Conceptualization, I.d.l.A.G. and M.A.G.; methodology, M.C.Z. and L.M.; software, I.d.l.A.G.; validation, I.d.l.A.G., M.A.G. and M.O.L.; formal analysis, I.d.l.A.G. and M.A.G.; investigation, I.d.l.A.G. and M.A.G.; resources, N.C.S.; data curation, I.d.l.A.G.; writing—original draft preparation, I.d.l.A.G.; writing—review and editing, M.A.G., M.C.Z. and L.M.; visualization, I.d.l.A.G.; supervision, N.C.S., M.C.Z. and L.M.; project administration, M.O.L. and N.C.S.; funding acquisition, M.O.L. and N.C.S. All authors have read and agreed to the published version of the manuscript.

Funding: This research received no external funding.

Institutional Review Board Statement: Not applicable.

Informed Consent Statement: Not applicable.

Data Availability Statement: Not applicable.

Acknowledgments: This work was supported by grant Ia ValSe-Food—CYTED (119 RT0567) and Secretaría de Ciencia y Técnica y Estudios Regionales—Universidad Nacional de Jujuy—CONICET, Argentina.

Conflicts of Interest: The authors declare no conflict of interest.

References

1. Xu, J.; Zhang, Y.; Wang, W.; Li, Y. Advanced properties of gluten-free cookies, cakes, and crackers: A review. *Trends Food Sci. Technol.* **2020**, *103*, 200–213. [CrossRef]
2. Nogueira, A.D.C.; Oliveira, R.A.D.; Steel, C.J. Protein enrichment of wheat flour doughs: Empirical rheology using protein hydrolysates. *Food Sci. Technol.* **2020**, *40*, 97–105. [CrossRef]
3. Gremasqui, I.D.L.A.; Giménez, M.A.; Lobo, M.O.; Sammán, N.C. Evaluation of functional and nutritional properties of hydrolyzed broad bean and quinoa flours. *Biol. Life Sci. Forum* **2021**, *8*, 12. [CrossRef]
4. Encina-Zelanda, C.; Teixeira, J.; Monteiro, F.C.; Gonzales-Barron, U.; Cadavez, V. Combined effect of xanthan gum on physico-chemical and textural properties of gluten-free batter and bread. *Food Res. Int.* **2018**, *111*, 544–555. [CrossRef]
5. Zúñiga-López, M.C.; Maturana, G.; Campmajó, G.; Saurina, J.; Nuñez, O. Determination of bioactive compounds in sequential extracts of chia leaf (*Salvia hipanica* L.) using UHPLC-HMRS (Q-Orbitrap) and a global evaluation of antioxidant in vitro capacity. *Antioxidant* **2021**, *10*, 1151. [CrossRef] [PubMed]
6. Minekus, M.; Alminger, M.; Alvito, P.; Balance, S.; Bohn, T.; Bourlieu, C.; Carriere, F.; Boutrou, R.; Corredig, M.; Dupont, D.; et al. A standardized static in vitro digestion method suitable for food-an international consensus. *Food Funct.* **2014**, *5*, 1113–1124. [CrossRef]
7. Ni, Q.; Ranawana, V.; Hayes, H.E.; Hayward, N.J.; Stead, D.; Raikos, V. Addition of broad bean hull to wheat flour for the development of high-fiber bread: Effects on physical and nutritional properties. *Foods* **2020**, *9*, 1192. [CrossRef]
8. Kulthe, A.A.; Thorat, S.S.; Lande, S.B. Evaluation of physical and textural properties of cookies prepared from pearl millet flour. *Int. J. Curr. Microbiol. Appl. Sci.* **2017**, *6*, 692–701. [CrossRef]
9. Batista, A.P.; Niccolai, A.; Fradinho, P.; Fragoso, S.; Bursic, I.; Rodolfi, L.; Biondi, N.; Tredici, M.R.; Sousa, I.; Raymundo, A. Microalgae biomass as an alternative ingredient in cookies: Sensory, physical and chemical properties, antioxidant activity and in vitro digestibility. *Algal Res.* **2017**, *26*, 161–171. [CrossRef]
10. Chinma, C.E.; Ibrahim, P.A.; Adedeji, O.E.; Ezeocha, V.C.; Ohuoba, E.U.; Kolo, S.I.; Abdulrahman, R.; Anumba Ogochukwu, E.U.; Adebo, J.A.; Adebo, O.A. Physicochemical properties, in vitro digestibility, antioxidant activity and consumer acceptability of biscuits prepared from germinated finger millet and Bambara groundnut flour blends. *Heliyon* **2022**, *8*, e10849. [CrossRef] [PubMed]
11. Mas, A.L.; Brigante, F.I.; Salvucci, E.; Ribotta, P.; Martinez, M.L.; Wunderlin, D.A.; Baroni, M.V. Novel cookie formulation with defatted sesame flour: Evaluation of its technological and sensory properties. Changes in phenolic profile, antioxidant activity, and gut microbiota after simulated gastrointestinal digestion. *Food Chem.* **2022**, *389*, 133122. [CrossRef]
12. Espinosa-Páez, E.; Hernández-Luna, C.E.; Longoria-García, S.; Martínez-Silva, P.A.; Ortiz-Rodríguez, I.; Villarreal-Vera, M.T.; Cantú-Saldaña, C.M. *Pleurotus ostreatus*: A potential concurrent biotransformation agent/ingredient on development of functional foods (cookies). *Lwt* **2021**, *148*, 111727. [CrossRef]
13. Bas, A.; El, S.N. Nutritional evaluation of biscuits enriched with cricket flour (*Acheta domesticus*). *Int. J. Gastron. Food Sci.* **2022**, *29*, 100583. [CrossRef]

biology and life sciences forum

Proceeding Paper

Effect of Pre-Treatment of Quinoa Seeds on Alcalase Hydrolysis and Antiradical Activity of Peptides Fractions [†]

Julio Rueda, Manuel O. Lobo and Norma C. Samman *

Centro Interdisciplinario de Tecnología y Desarrollo Social, CIITED-UNJu-CONCIET, Universidad Nacional de Jujuy, Ítalo Palanca 10, San Salvador de Jujuy 4600, Argentina; julioruedafca@gmail.com (J.R.); mlobo958@gmail.com (M.O.L.)

* Correspondence: normasamman@gmail.com

[†] Presented at the V International Conference la ValSe-Food and VIII Symposium Chia-Link, Valencia, Spain, 4–6 October 2023.

Abstract: The pre-treatment of seeds prior to processing is gaining attention as alternative ways of modifying properties of foods. In this work, the extrusion and germination of quinoa seeds were evaluated for their effect on protein profile, alcalase hydrolysis and the antioxidant capacity (AOC) of peptide fractions (>10, 3–10 and <3 kDa). The proteins in extruded (EQ), germinated (GQ) and unprocessed (UQ) seeds were extracted, hydrolysed and fractionated by ultrafiltration. An SDS-PAGE protein profile showed that the pre-treatments partially hydrolysed high-molecular-weight proteins (75–100 kDa) into low-molecular-weight polypeptides, and chenopodin was unaltered. The hydrolysis degree in hydrolysates reached 38.38% for UQ seeds, 31.85% for EQ seeds and 30.09% for GQ seeds. Compared to the UQ hydrolysate, the extrusion and germination significantly improved ($p < 0.05$) the AOC of the <3 kDa fraction by 61.49% (EQ) and 38.11% (GQ) and the 3–103–10 kDa fraction by 130.98% (EQ) and 57.71% (GQ). The pre-treatment of seeds before protein extraction and hydrolysis modified the peptide profile with improved antiradical activity after alcalase hydrolysis. This study highlights the use of mild pre-treatments applied to quinoa seeds as a way to modify proteins and obtain hydrolysates with enhanced bioactivity.

Keywords: antioxidant; extrusion; hydrolysate; germination; quinoa; whole seeds

Citation: Rueda, J.; Lobo, M.O.; Samman, N.C. Effect of Pre-Treatment of Quinoa Seeds on Alcalase Hydrolysis and Antiradical Activity of Peptides Fractions. *Biol. Life Sci. Forum* **2023**, *25*, 8. https://doi.org/10.3390/blsf2023025008

Academic Editors: Claudia M. Haros, Loreto Muñoz and María Dolores Ortolá

Published: 28 September 2023

1. Introduction

The interest in environmentally safe processes such as extrusion, germination, enzymatic hydrolysis, etc., has increased because of their reduced negative impact on the physical, chemical and biological properties of foodstuffs [1]. Extrusion denatures proteins, improves enzymatic hydrolysis and modifies amino acid profiles [2]. With germination, storage proteins break down, and the profile of essential and non-essential amino acids changes [3]. Food processing technologies such as extrusion and germination can improve the properties of protein hydrolysates. An improved AOC was reported for the hydrolysates of extruded [4] and germinated seeds [5]. However, Montoya-Rodríguez et al. [6] reported that the AOC diminished after the extrusion of amaranth seeds. Hydrolysate production follows the general sequence of protein extraction, enzymatic hydrolysis, fractionation and testing of bio-functional properties; however, few reports describe the effect of pre-treatments applied to seeds before protein extraction. The worldwide-recognised Andean seed quinoa provides all the essentials and a balanced amino acid profile, making it a unique protein among its counterparts and other cereals [7]. In this work, the extrusion and germination processes were evaluated as pre-treatments of quinoa seeds for their potential effect on the protein profile, the hydrolysis performance, and the AOC of peptides.

2. Materials and Methods

2.1. Quinoa Seeds Origin

Desaponified Quinoa (*Chenopodium quinoa* Wild) seeds (var. Hornillos-Gob Jujuy INTA) were purchased at the National Institute of Agricultural Technologies (INTA-IPAF) (Maimará, Jujuy, Northwest Argentina).

2.2. Quinoa Seeds Processing

The extrusion of seeds was carried out using a twin-screw extruder (INCALFER, DT65, Buenos Aires, Argentina). The extrusion conditions were humidity of 28% (*w/w*), temperature of three sections extruder barrel at 45, 90, and 125 °C and frequency of 8.7 (screw), 12.1 (feeding), and 17.1 (cutter) Hertz. The final products were aired and stored at 4 °C in sealed polyethylene bags. The germination of disinfected (0.5 kg; sodium hypochlorite 100 ppm; 10 min) and washed seeds (3 times, distilled water) was conducted over tissue paper, and incubation was carried out(30 °C, 24 h, humidity >90%,) in the dark (Memmert Radiant Warmer Model A52200-35-Vac 230, Büchenbach, Germany). Germinated seeds (max. 2.5 mm radicle) were dried (~10% moisture content) in an air-drying oven (40 °C).

2.3. Protein Concentrate Preparation

Milled flour from germinated, extruded and unprocessed seeds was defatted (1:5; in petroleum ether), and protein concentrates (PCs) were prepared by alkaline solubilization followed by centrifugation and acid precipitation [8]. Polyphenols were removed with anhydrous ethanol, and the PC was neutralised to a pH of 7 (NaOH) and dried at 30 °C. Total nitrogen was determined via the Kjeldahl method.

2.4. Electrophoretic Profile by SDS-PAGE

The samples of protein concentrates were solubilised (pH 8) and boiled (3 min) in running buffer (2% SDS), glycerol (10%), bromophenol blue (0.01%), Tris-HCl buffer (pH 6.8) and mercaptoethanol (5%), and 5 μL of them were put in Laemmli buffer and stacking (60 V) and running gels (120 V) (Bio-Rad Laboratories, Hercules, CA, USA). Gels were dyed (Coomassie R-250) and decoloured (methanol/acetic acid/water (50/20/30) [8]. Bands were identified (GelAnalizer 19.1. 2010) with a standard protein (6.5–200 kDa, Sigma-Aldrich, Steinheim, Germany).

2.5. Protein Hydrolysis, Fractionation and Quantification of Peptides

Each protein (1% *w/v*; 300 mL) was hydrolysed with alcalase (E:S 1:10) at a constant pH of 9, 50 °C and a NaOH concentration of 0.3 M. Enzyme inactivation was at 95 °C for 10 min. The hydrolysis degree (HD%) was calculated via the pH-stat method. The peptide fractions (Amicon® 10 and 3 MWCO centrifugal filters, Millipore, Burlington, MA, USA) were lyophilised, mixed with phosphate buffer (pH 8.2, 3.4 mL, 0.2 M) and TNBS (0.5 mL, 5% *v/v*), incubated (dark, 50 °C, 60 min, 200 rpm) for peptide quantification at 420 nm and plotted against a standard (L-leucine, mM/mL) [8].

2.6. Antiradical Activity of Peptide Fractions

Radical ABTS [2,2′-azino-bis (3-ethylbenzothiazoline-6-sulfonic acid)] was formed for 12–16 h in the dark (7 mM in water and ammonium persulphate 2.45 mM). Working radical solutions (0.7 absorbance, 734 nm) were blended with peptide fractions (10 μL, 1 mg/mL), and the absorbance was measured after 6 min. Ascorbic acid was plotted as the standard [8].

2.7. Statistics Analysis

Data are expressed as mean ± standard deviation (SD) of three replicates. Means were compared by two-way analysis of variance (ANOVA), and differences were compared with the Tukey test ($p < 0.05$). GraphPad Prism® V5.03 software (GraphPad Software Inc., San Diego, CA, USA) was used for graph plotting.

3. Results and Discussion

3.1. Electrophoretic Protein Profile of Extruded, Geminated and Control Quinoa Proteins

The electrophoretic profile of the protein concentrates of unprocessed (UQ), extruded (EQ), and germinated (GQ) seeds is show in Figure 1. The lanes UQ, EQ and GQ displayed bands of MW (kDa) of 45.70 (A), 29.70–25.10 (B) and 15.93–20.00 (C). Bands B and C are comparable to chenopodin, which is found in quinoa [9]. After processing, the intensity in the 93–79 kDa band (lane UQ) was reduced by 70.11± 15% (EQ) and by 42.74 ± 19% (GQ). This reduced band intensity may be due to denaturation, proteolysis or insoluble aggregates. Furthermore, the EQ and GQ lanes augmented their intensity by 140.76 ± 12% (EQ) and 107.21 ± 14% (GQ) for the <14.2 kDa band, suggesting a higher concentration of polypeptides. Similar results were also described in sprouted quinoa, amaranth [9,10] and extruded amaranth seeds [6]. This evidence demonstrates that extrusion and germination partially hydrolysed high-MW proteins and formed low-MW polypeptides. Each protein concentrate showed different protein patterns and polypeptide profiles.

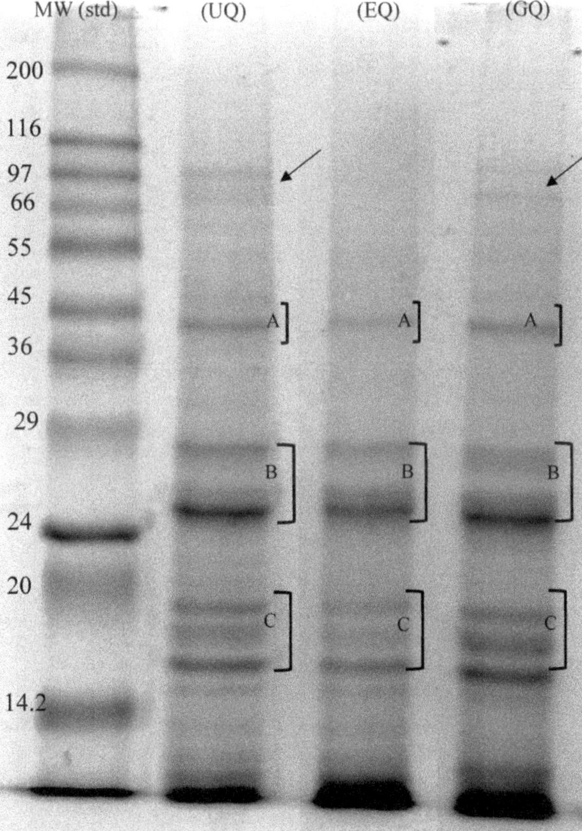

Figure 1. Electrophoretic profile (SDS-PAGE) of alkaline soluble quinoa proteins. Unprocessed (UQ), extruded (EQ) and germinated (GQ) seeds protein concentrate. Molecular weight standard (MW std). MW of chenopodin 45.70 (**A**), 29.70–25.10 acid (**B**) and 15.93–12.54 basic (**C**), subunits. Black arrows: high molecular weight proteins (HMW proteins).

3.2. Impact of Processing on the Degree of Hydrolysis

Table 1 shows that the HD% increased noticeably (30 min), reaching a significantly higher value for the UQPH (28.45%) compared to the GPH (23.98%) ($p < 0.05$) and the EPH (25.04%). At the end of the hydrolysis (150 min), the HD% was 35.38 (UQPH), 31.85 (EPH) and 30.09% (GPH). The differences in the HD% could be attributed to the presence of more proteins susceptible to hydrolysis in the UQPH than in the EPH and the GPH, as shown in the electrophoretic run lane UQ (Figure 1). The low HD% in EPH could also be attributed to less soluble proteins or aggregates being formed during extrusion. Similar conclusions were reported by Montoya-Rodríguez et al. [6] for extruded amaranth seeds. However, an increase in the HD% was reported for pea (*Pisum sativum*) extruded protein [4]. According to Nor Afizah and Rizvi [11], an extrusion at <75 °C forms aggregates susceptible to hydrolysis, but and extrusion at >90 °C forms aggregates resistant to hydrolysis. In the case of germination, the internal proteolysis caused the degradation of high-MW proteins and the formation of short polypeptides (Figure 1, lane GQ). Similar results have been reported for germinated quinoa [12].

Table 1. Hydrolysis degree kinetics (HD%) of protein concentrates from unprocessed, extruded and germinated quinoa seeds.

Time (min)	UQPH	GPH	EPH
0	n.a	n.a	n.a
30	28.45 [a] ± 1.00	23.98 [b] ± 3.32	25.04 [ab] ± 0.83
60	31.27 [a] ± 1.00	27.74 [a] ± 2.66	28.80 [a] ± 0.83
120	34.56 [a] ± 1.00	27.51 [b] ± 1.00	30.91 [ab] ± 0.83
150	35.38 [a] ± 0.83	30.09 [b] ± 3.32	31.85 [ab] ± 0.83

UQPH: unprocessed seed protein hydrolysate; GPH: germinated seed protein hydrolysate; EPH: extruded seed protein hydrolysate. Data are represented as mean ± standard deviation (n = 3); different letters in the same row indicate a statistically significant difference ($p < 0.05$). n.a: not applicable.

3.3. Peptide Quantification in Protein Hydrolysates

Generally, bioactive peptides are composed for up to 20 amino acids. Figure 2 shows the distribution of peptides in the >10, 3–10 and <3 kDa fractions. The largest abundance of peptides was found in the <3 kDa fraction for the UQPH, EQPH and GQPH, followed by the 3–10 and >10 kDa fractions. The extrusion significantly augmented ($p < 0.05$) the concentration of peptides in all fractions, which may be attributed to the formation of polypeptides, as shown by the SDS-PAGE profile (Figure 1 lane EQ). Similar findings were reported for amaranth seeds [6]. These data suggest that a two-step hydrolysis occurred: the first one during the extrusion, followed by the enzymatic hydrolysis per se.

3.4. AOC of Peptide Fractions

The contribution of each peptide fraction to the total AOC of the hydrolysate was studied. Figure 3 shows that all fractions displayed antiradical activity. The <3 kDa fraction had the highest AOC in the UQPH, EPH and GPH. The <3 kDa fraction in the EPH and GPH displayed increases ($p < 0.05$) of 61.49% and 38.11%. The 3–10 kDa fraction in the EPH and GPH showed improvements ($p < 0.05$) by 130.98% and 57.71%, compared to UQPH. Although the UQPH and GPH had similar concentrations of peptides in the 3–10 kDa fraction(Figure 2), the GPH had a higher AOC, which suggests the presence of different antiradical peptides. The EPH and GPH displayed a similar HD%, but this was lower than that of the UQPH (Table 1), which suggests that the better AOC could also be related to the presence of large peptides. Overall, the presented data suggest that not only short peptides (<3 kDa) exhibited AOC but also polypeptides between 10 and 3 kDa, due to extrusion and germination. The improvement in the AOC could be attributed to changes in the amino acid composition, peptide profile and steric factors, as well as the production of new, larger peptides due to the extrusion and germination of seeds. Other authors have reported similar results for extruded proteins and peptides [4,13]. Additionally, the type of enzyme and the hydrolysis time was also reported to influence the AOC [6]. Similarly, Montoya-Rodríguez et al. [6] found an enhanced AOC of hydrolysates from extruded amaranth seeds and for germinated amaranth seeds [10].

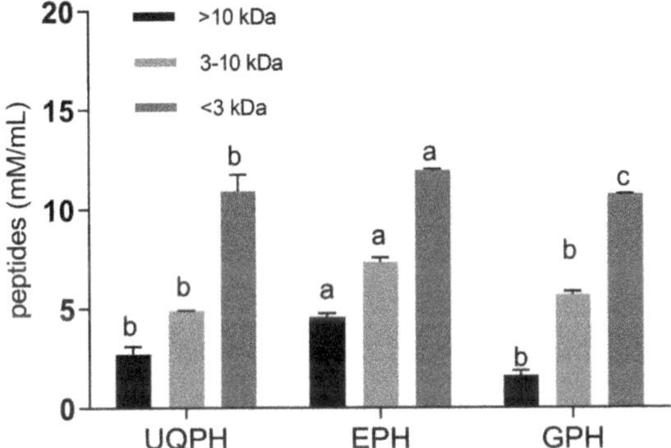

Figure 2. Abundance of peptides (leucine mM/mL) for different MW fractions in protein hydrolysates from unprocessed quinoa seed protein hydrolysate (UQPH), extruded seed protein hydrolysate (EPH) and germinated seed protein hydrolysate (GPH) after 180 min of hydrolysis. Data are shown as mean ± standard deviation (n = 3). Different letters in the same molecular weight fraction indicate a statistically significant difference ($p < 0.05$).

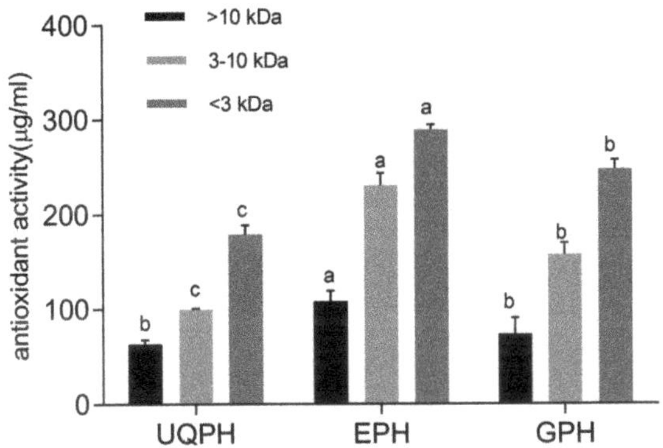

Figure 3. Antioxidant activity (ascorbic acid equivalents µg/mL) of hydrolysate fractions from unprocessed quinoa seed protein hydrolysate (UQPH), extruded seed protein hydrolysate (EPH) and germinated seed protein hydrolysate (GPH) after 180 min hydrolysis. Data shown as mean ± standard deviation (n = 3). Different letters in same molecular weight fraction indicate statistically significant difference ($p < 0.05$).

4. Conclusions

The novel Andean Argentine quinoa variety, Hornillos-Gob Jujuy INTA, was suitable for extrusion and germination for producing PCs with considerable high-protein purity (61–66%). The processing of the seeds did not alter the chenopodin, but the HMW proteins were partially hydrolysed. Distinctive protein hydrolysates were prepared, and the proteolysis kinetics were modulated by reducing the HD. The pre-treatments that were applied enhanced the AOC of the long (3–10 kDa) and short (<3 kDa) peptides. The improvement may be due to the synergistic effects of different peptides and the new peptides formed due to processing. These results provide an insight into how affordable, green and controlled pre-treatments of quinoa seeds may change and modulate the hydrolysis products with enhanced bio-functional properties.

Author Contributions: Conceptualization, N.C.S.; methodology, J.R. and M.O.L.; software, J.R.; validation, M.O.L. and N.C.S.; formal analysis, J.R. and M.O.L.; investigation, J.R., M.O.L. and N.C.S.; resources, N.C.S. and M.O.L.; data curation, M.O.L. and N.C.S.; writing—original draft preparation, J.R.; writing—review and editing, M.O.L. and N.C.S.; visualization, J.R.; supervision, M.O.L. and N.C.S.; project administration, J.R.; funding acquisition, M.O.L. and N.C.S. All authors have read and agreed to the published version of the manuscript.

Funding: This work was supported by grant Ia ValSe-Food-CYTED (119RT0567), Secretaría de Ciencia, Técnica y Estudios Regionales, Universidad Nacional de Jujuy and CONICET.

Institutional Review Board Statement: Not applicable.

Informed Consent Statement: Not applicable.

Data Availability Statement: Not applicable.

Acknowledgments: The authors thank to Ia ValSe-Food-CYTED network (119RT0567), Secretaría de Ciencia, Técnica y Estudios Regionales, Universidad Nacional de Jujuy (UNJu-Argentina) and Scientific and Technological Research Council (CONICET-Argentina).

Conflicts of Interest: The authors declare no conflict of interest.

References

1. López-Pedrouso, M.; Díaz-Reinoso, B.; Lorenzo, J.M.; Cravotto, G.; Barba, F.J.; Moure, A.; Domínguez, H.D.F. Applications of green technologies-based approaches for food safety enhancement: A comprehensive review. *Food Sci. Nutr.* **2022**, *10*, 2855–2867. [CrossRef]
2. Kamau EH Nkhata, S.G.; Ayua, E.O. Extrusion and nixtamalization conditions influence the magnitude of change in the nutrients and bioactive components of cereals and legumes. *Food Sci. Nutr.* **2020**, *8*, 1753–1765. [CrossRef] [PubMed]
3. Uppal, V.; Bains, K. Effect of germination periods and hydrothermal treatments on in vitro protein and starch digestibility of germinated legumes. *J. Food Sci. Technol.* **2012**, *49*, 184–191. [CrossRef] [PubMed]
4. Zhou, Q.C.; Liu, N.; Feng, C.X. Research on the effect of papain co-extrusion on pea protein and enzymolysis antioxidant peptides. *J. Food Process. Preserv.* **2017**, *41*, e13301. [CrossRef]
5. de Souza Rocha, T.; Hernandez, L.M.R.; Mojica, L.; Johnson, M.H.; Chang, Y.K.; González de Mejía, E. Germination of Phaseolus vulgaris and alcalase hydrolysis of its proteins produced bioactive peptides capable of improving markers related to type-2 diabetes in vitro. *Food Res. Int.* **2015**, *76*, 150–159. [CrossRef]
6. Montoya-Rodríguez, A.; de Mejía, E.G.; Dia, V.P.; Reyes-Moreno, C.; Milán-Carrillo, J. Extrusion improved the anti-inflammatory effect of amaranth (Amaranthus hypochondriacus) hydrolysates in LPS-induced human THP-1 macrophage-like and mouse RAW 264.7 macrophages by preventing activation of NF-κB signaling. *Mol. Nutr. Food Res.* **2014**, *58*, 1028–1041. [CrossRef] [PubMed]
7. Filho, A.M.M.; Pirozi, M.R.; Borges, J.T.D.S.; Pinheiro Sant'Ana, H.M.; Chaves, J.B.P.; Coimbra, J.S.D.R. Quinoa: Nutritional, functional, and antinutritional aspects. *Crit. Rev. Food Sci. Nutr.* **2017**, *57*, 1618–1630. [CrossRef]
8. Rueda, J.; Lobo, M.O.; Sammán, N. Changes in the Antioxidant Activity of Peptides Released during the Hydrolysis of Quinoa (Chenopodium quinoa willd) Protein Concentrate. *Proceedings* **2020**, *53*, 12. [CrossRef]
9. Mäkinen, O.E.; Hager, A.S.; Arendt, E.K. Localisation and development of proteolytic activities in quinoa (Chenopodium quinoa) seeds during germination and early seedling growth. *J. Cereal Sci.* **2014**, *60*, 484–489. [CrossRef]
10. Sandoval-Sicairos, E.S.; Milán-Noris, A.K.; Luna-Vital, D.A.; Milán-Carrillo, J.; Montoya-Rodríguez, A. Anti-inflammatory and antioxidant effects of peptides released from germinated amaranth during in vitro simulated gastrointestinal digestion. *Food Chem.* **2021**, *343*, 128394. [CrossRef] [PubMed]
11. Nor Afizah, M.; Rizvi, S.S.H. Functional properties of whey protein concentrate texturized at acidic pH: Effect of extrusion temperature. *Lwt* **2014**, *57*, 290–298. [CrossRef]
12. Obaroakpo, J.U.; Liu, L.; Zhang, S.; Lu, J.; Pang, X.; Lv, J. α-Glucosidase and ACE dual inhibitory protein hydrolysates and peptide fractions of sprouted quinoa yoghurt beverages inoculated with Lactobacillus casei. *Food Chem.* **2019**, *299*, 124985. [CrossRef] [PubMed]
13. Yuan, G.; Pan, Y.; Li, W.; Wang, C.; Chen, H. Effect of extrusion on physicochemical properties, functional properties and antioxidant activities of shrimp shell wastes protein. *Int. J. Biol. Macromol.* **2019**, *136*, 1096–1105. [CrossRef] [PubMed]

Proceeding Paper

Mistol-Based Vegan Beverages for a Healthy Diet [†]

R. Villalba, J. Belotto, E. Coronel, A. Suárez, S. Caballero and L. Mereles *

Departamento Bioquímica de Alimentos, Facultad de Ciencias Químicas, Universidad Nacional de Asunción, San Lorenzo P.O. Box 1055, Paraguay; rvillalba@qui.una.py (R.V.); jbelotto@qui.una.py (J.B.); ecoronel@qui.una.py (E.C.); adecia.suarez@gmail.com (A.S.); scaballero@qui.una.py (S.C.)

* Correspondence: lauramereles@qui.una.py

[†] Presented at the V International Conference la ValSe-Food and VIII Symposium Chia-Link, Valencia, Spain, 4–6 October 2023.

Abstract: The fruit of *Sarcomphalus mistol* (Griseb.) is from the Great American Chaco, an important wild food resource. However, its composition and potential uses are unknown to a large part of the population. In this work, two mistol and peanuts-based products were prepared (a toasted mistol product "mistol coffee" and a mistol-peanuts-based vegan beverage), and its nutritional value and antioxidant potential was investigated. The chemical composition, total phenolics compounds (TPCs), and total antioxidant capacity (TAC) of the two products were analyzed on a dry basis with AOAC, Folin–Ciocalteu, and ABTS+ radical inhibition methods, respectively. The toasted mistol product was high in dietary fiber (51.89 ± 0.65 g/100 g) and TPC (2729 ± 362 mgGAE/100 g), which was consistent with the observed high TAC (307 ± 14 mM TEAC/g). Minerals found in the toasted mistol product were mainly calcium and magnesium. The mistol-peanut-based vegan beverage presented 12.67 g/100 g of dry extract, and its main nutritional contributions are carbohydrates (8.16 ± 1.05 g/100 g db), lipids (2.33 ± 0.29 g/100 g db), magnesium (10.34 ± 1.40 mg/100 g db), and polyphenols (345 ± 4.58 mg GAE/100 g db) in agreement with their TAC in the lyophilized product (17.6 ± 5.25 mM/TEAC/g db). These products can be used according to their observed nutritional value and antioxidant potential as foods with healthy properties, especially for vegan populations.

Keywords: mistol; peanuts; vegan; nutritive beverages; antioxidants; healthy diets

Citation: Villalba, R.; Belotto, J.; Coronel, E.; Suárez, A.; Caballero, S.; Mereles, L. Mistol-Based Vegan Beverages for a Healthy Diet. *Biol. Life Sci. Forum* **2023**, 25, 9. https://doi.org/10.3390/blsf2023025009

Academic Editors: Claudia M. Haros, Loreto Muñoz and María Dolores Ortolá

Published: 28 September 2023

1. Introduction

Diet drinks are recognized as sources of energy and nutrients, but may also provide health benefits due to the presence of bioactive compounds [1]. The fruit of mistol (*Sarcomphalus mistol* Griseb.) is a drupe with pasty and sweet pulp, with exquisite flavor and reddish-hazelnut color from the great American Chaco, which is appreciated by children in indigenous communities. However, its composition is unknown to a large part of the population [2]. It is consumed directly but is also used as an ingredient in other food preparations (such as "bolanchao"), as an infusion and a toasted product ("mistol coffee"). The tea has medicinal properties against biliary colic, dysentery, cough, heart diseases, hypertension, diabetes, as an antidote against snake bites and poisonous insects, and as a natural energizer [3]. They can provide 8% protein (28.3% essential amino acids of the total AA contents). Their extracts enriched in polyphenols are important source of flavonoids with antioxidant capacity, to inhibit α-glucosidase, α-amylase, and pancreatic lipase, enzymes that are related to metabolic syndrome [4]. Peanuts (*Arachis hypogaea* L.) are the most widely cultivated grain legume and are an important source of protein and oils in many countries; the content of unsaturated fatty acids is mainly in the oil (83–87%) [5].

Plant-based nutritional drinks have good future prospects, especially with the use of natural ingredients containing low sugar or with no-added-sugar drinks, rich in polyphenols, and with antioxidant properties, which may also confer additional health benefits [5].

The aim of this work was to describe the nutritional value of a mistol-peanut vegan beverage and toasted mistol product with antioxidant potential, as an alternative nutritional drink based on vegetables and low in sugar for a healthy diet.

2. Materials and Methods

Two mistol-based products were prepared: a toasted mistol product "mistol coffee" and a mistol-peanut-based vegan beverage prepared in triplicate in a laboratory. Wild mistol fruits and cultivated peanuts were harvested from Philadelphia, Chaco. The ripe whole mistol fruits were washed and dried (in the sun for 3 days), roasted over a dry heat in a pan, and ground according to the traditional procedure. The mistol-peanut-based vegan beverage was produced from the infusion of the 3% toasted mistol product substitutes in an extraction time of 1 minute. The drink was obtained by mixing 50% toasted mistol infusion with 50% "peanut milk" to which 0.3% lecithin was added as an emulsifier and 7% organic sugar.

The samples of the roasted mistol product and the lyophilized drink have been analyzed using official methods of the AOAC and AOCS [6]: moisture (AOCS Ca 2b-08), ash (AOAC 900.2), total lipids (AOAC 948.22), dietary fiber (AOAC 985.29), protein (AOAC 979.09), and total carbohydrates and sugars using the anthrone method [7]. The caloric value was determined using the Atwater method. The mineral elements sodium, calcium, magnesium, manganese, zinc, iron, and copper were determined using the AOAC 975.03 method, and phosphorus using the AOAC 970.39 spectrophotometric method. The TPC was determined using the Folin–Ciocalteu oxidation-reduction spectrophotometric method [8], and the TAC using the ABTS+ radical inhibition method [9]. The data were collected and processed using the GraphPad Prism 8.2 program (San Diego, CA, USA). To determine significant differences, Student's test was applied with a value of $p \leq 0.05$ considered significant.

3. Results

Toasted mistol product power was rich in dietary fiber (51.89 ± 0.65 g/100 g) and polyphenols (2729 ± 362 mgGAE/100 g), which was consistent with the high TAC observed (307 ± 14 mM TEAC/g). The main minerals found were calcium and magnesium (Table 1). The mistol-peanut-based vegan beverage lyophilized presented higher total carbohydrates content, lipids, magnesium, and polyphenols (547 ± 95 mg GAE/100 g) in agreement with their TAC (17.6 ± 5.25 mM/TEAC/g). The dry extract of mistol with peanuts presented a higher carbohydrate content (64.40 ± 4.35 g/100 g) than the mistol coffee (21.91 ± 0.65 g/100 g), composed mostly of simple sugars. The lipids detected (18.38 g/100 g) come from the peanuts used as a raw material to make the drink. The total polyphenol content in the toasted mistol was significantly higher (*t* test, $p < 0.05$). Both products presented great antioxidant capacity on a dry basis (Table 1).

Table 1. Chemical composition of toasted mistol product and lyophilized mistol-peanuts-based vegan beverage on a dry basis.

	Toasted Mistol Product	Mistol-Peanuts-Based Vegan Beverage
	Centesimal composition	
Moisture (g/100 g)	3.23 ± 0.27 [a]	1.60 ± 0.13 [b]
Protein (g/100 g)	13.00 ± 1.3 [a]	11.29 ± 0.31 [a]
Total carbohydrates (g/100 g)	21.91 ± 0.65 [a]	64.40 ± 4.35 [b]
Sugars (g/100 g)	19.53 ± 0.39 [a]	61.4 ± 0.68 [b]
Total lipids (g/100 g)	Tz	18.38 ± 0.49 [b]
Ash (g/100 g)	5.69 ± 0.12 [a]	1.51 ± 0.06 [b]
Dietary fiber (g/100 g)	46.64 ± 0.65 [a]	3.19 ± 0.85 [b]
Caloric value (Kcal/100 g)	140 ± 9 [a]	468 ± 21 [b]

Table 1. *Cont.*

	Toasted Mistol Product	Mistol-Peanuts-Based Vegan Beverage
Minerals		
Calcium (mg/100 g)	223 ± 4 [a]	21.74 ± 1.97 [b]
Phosphorus (mg/100 g)	174 ± 5 [a]	182 ± 2 [a]
Magnesium (mg/100 g)	95.77 ± 7.02 [a]	81.58 ± 6.40 [a]
Iron (mg/100 g)	7.83 ± 0.22 [a]	1.85 ± 0.7 [b]
Sodium (mg/100 g)	27.58 ± 4.51 [a]	27.23 ± 3.89 [a]
Zinc (mg/100 g)	0.95 ± 0.76 [a]	0.74 ± 0.15 [a]
Copper (mg/100 g)	0.38 ± 0.01 [a]	0.53 ± 0.02 [b]
Manganese (mg/100 g)	<0.010	0.10 ± 0.07 [a]
Antioxidants		
TPC (mgGAE/100 g)	2729 ± 362 [a]	547 ± 94 [b]
ABTS (mM TEAC/g)	307 ± 14 [a]	17.6 ± 5.25 [b]

Results are expressed as the mean ± standard deviation of three independent tests. Values with the same letters indicate that there is no significant difference (Student's *t* test, $p < 0.05$). Tz: traces.

4. Discussion

The results show a potential nutritional contribution of the formulated beverage with mistol and peanuts with a dry extract of 12.67 g/100 g. The contribution of carbohydrates and lipids in the reconstituted drink was 8.16 ± 1.05 and 2.33 ± 0.29 g/100 g, respectively. A serving of drink is considered to be 200 mL, which in addition to being a nutritious drink (16.32 g of carbohydrates), would provide antioxidant compounds and a low sodium content (6.9 mg/serving). One of the most consumed vegetable drinks is soy milk, which provides 6–12% carbohydrates in a 200 mL portion. On the other hand, the recommendation is to choose fortified drinks that contain at least 4.8 g of protein per 200 mL. However, we observe that the drink prepared in this study provides a much higher sugar content (16.32 g/200 mL) and lower protein (2.86/200 mL per serving) compared to soy milk [10]. Regarding minerals, a serving of drink provides 15% of the recommended daily intake of copper. The information provided here is relevant for the development of new products, the commercialization and promotion of the consumption of foods derived from native resources, within the framework of food security and environmental sustainability towards better nutrition. The mistol and peanuts seeds are important foods with great nutritional potential in the regional diet as ingredients of beverages. Extremely high levels of resistance or immunity make wild *Arachis* species a highly valuable genetic resource [11]. The mistol tree is a deciduous honey plant that also provides a significant contribution to the environment by fixing carbon and protecting soils and biodiversity [3]. Peanut species and their varieties, as well as mistol trees, are of great importance for conservation as resources for food security.

5. Conclusions

Regional food resources, such as mistol fruits and peanut seeds, which are rarely used as ingredients in beverages, would be valuable for the development of this type of plant-based nutritional product, with an important contribution of macro and micronutrients as well as polyphenols with recognized properties and antioxidants, especially the roasted mistol product called "mistol coffee" and peanut-based formulations.

Author Contributions: Conceptualization, L.M. and R.V.; methodology, S.C. and L.M.; software, E.C.; validation, J.B. and L.M.; formal analysis, R.V., A.S. and J.B.; investigation, R.V.; resources, S.C.; data curation, L.M.; writing—original draft preparation, L.M.; writing—review and editing, L.M.; visualization, J.B. and R.V.; supervision, L.M.; project administration, L.M. All authors have read and agreed to the published version of the manuscript.

Funding: This work was supported by grant Ia ValSe-Food-CYTED (119RT0567), COOPI (EuropeAid/154653/DD/ACT/Multi, UE), and Adeline Friesen -Tucos Factory E.I.R.L., Paraguay.

Institutional Review Board Statement: Not applicable.

Informed Consent Statement: Not applicable.

Data Availability Statement: Not applicable.

Acknowledgments: Ia ValSe-Food-CYTED (119RT0567), COOPI and Adeline Friesen.

Conflicts of Interest: The authors declare no conflict of interest.

References

1. Lodhi, S.; Vadnere, G.P. Health-Promoting Ingredients in Beverages. In *Value-Added Ingredients and Enrichments of Beverages*; Elsevier: Amsterdam, The Netherlands, 2019; pp. 37–61; ISBN 978-0-12-816687-1.
2. Cardozo, M.L.; Ordoñez, R.M.; Alberto, M.R.; Zampini, I.C.; Isla, M.I. Antioxidant and Anti-Inflammatory Activity Characterization and Genotoxicity Evaluation of Ziziphus Mistol Ripe Berries, Exotic Argentinean Fruit. *Food Res. Int.* **2011**, *44*, 2063–2071. [CrossRef]
3. Cittadini, M.C.; García-Estévez, I.; Escribano-Bailón, M.T.; Bodoira, R.M.; Barrionuevo, D.; Maestri, D. Nutritional and Nutraceutical Compounds of Fruits from Native Trees (*Ziziphus mistol* and *Geoffroea decorticans*) of the Dry Chaco Forest. *J. Food Compos. Anal.* **2021**, *97*, 103775. [CrossRef]
4. Orqueda, M.E.; Zampini, I.C.; Torres, S.; Alberto, M.R.; Pino Ramos, L.L.; Schmeda-Hirschmann, G.; Isla, M.I. Chemical and Functional Characterization of Skin, Pulp and Seed Powder from the Argentine Native Fruit Mistol (*Ziziphus mistol*). Effects of Phenolic Fractions on Key Enzymes Involved in Metabolic Syndrome and Oxidative Stress. *J. Funct. Foods* **2017**, *37*, 531–540. [CrossRef]
5. Zahran, H.A.; Tawfeuk, H.Z. Physicochemical Properties of New Peanut (*Arachis hypogaea* L.) Varieties. *OCL* **2019**, *26*, 19. [CrossRef]
6. Horwitz, W.; Chichilo, P.; Reynols, H. *Official Methods of Analysis of the Association of Official Analytical Chemists*, 17th ed.; AOAC: Gaithersburg, MA, USA, 2000.
7. Dreywood, R. Qualitative Test for Carbohydrate Material. *Ind. Eng. Chem. Anal. Ed.* **1946**, *18*, 499. [CrossRef]
8. Singleton, V.L.; Orthofer, R.; Lamuela-Raventós, R.M. [14] Analysis of total phenols and other oxidation substrates and antioxidants by means of folin-ciocalteu reagent. *Methods Enzymol.* **1999**, *299*, 152–178.
9. Re, R.; Pellegrini, N.; Proteggente, A.; Pannala, A.; Yang, M.; Rice-Evans, C. Antioxidant Activity Applying an Improved ABTS Radical Cation Decolorization Assay. *Free Radic. Biol. Med.* **1999**, *26*, 1231–1237. [CrossRef] [PubMed]
10. Sethi, S.; Tyagi, S.K.; Anurag, R.K. Plant-Based Milk Alternatives an Emerging Segment of Functional Beverages: A Review. *J. Food Sci. Technol.* **2016**, *53*, 3408–3423. [CrossRef] [PubMed]
11. De Egea Elsam, J.; Céspedes, G.; Peña-Chocarro, M.d.C.; Mereles, F.; Rolón Mendoza, C. *Recursos Fitogenéticos del Paraguay: Sinopsis, Atlas y Estado de Conservación de los Parientes Silvestres de Especies de Importancia para la Alimentación y la Agricultura (Parte I)*; Universidad Autónoma de Asunción (UAA): Asunción, Paraguay, 2018; p. 232.

biology and life sciences forum

Proceeding Paper

Use of By-Products of Selection Process of Bean (*Phaseolus vulgaris* L.): Extraction of Protein and Starch [†]

Matias M. Alancay, Sonia R. Calliope, Rita M. Miranda and Norma C. Samman *

Centro Interdisciplinario de Investigaciones en Tecnologías y Desarrollo Social para el NOA (CIITED, UNJu-CONICET), Facultad de Ingeniería, Universidad Nacional de Jujuy, San Salvador de Jujuy 4600, Argentina; matiasalancay@yahoo.com (M.M.A.); scalliope@ciited.edu.edu.ar (S.R.C.); rmiranda@unju.edu.ar (R.M.M.)

* Correspondence: normasamman@gmail.com

[†] Presented at the V International Conference la ValSe-Food and VIII Symposium Chia-Link, Valencia, Spain, 4–6 October 2023.

Abstract: The industrial selection process of bean (*Phaseolus vulgaris* L.) produced in Northwest Argentina (NOA) region produces 7000 tons/year of by-products integrated from broken, bruised, and reduced-sized seeds. This investigation aimed to study the possibilities of using these by-products as a source of protein and starch. Samples were crushed to obtain flour (BF) with particle size of 250 μm. Starch and protein were extracted in a 6:1 and 10:1 water/flour ratio at pH 9 and 10, respectively. After centrifugation, the protein was precipitated from the supernatant at pH 4.5, and a bean protein concentrate (BPC) was obtained. The chemical composition of BF, S, and BPC was determined. Starch swelling power (SP), water solubility index (WSI), water absorption index (WAI), and syneresis in cooling (SC) and freezing (SF) conditions were determined. The proportion of molecular structure of BPC was determined using deconvolution of infrared spectrum (Amide I zone), and their solubility using Bradford reactive. The yield of obtaining processes of BPC and bean starch (BS) of high purities was 13.0 and 50.3 g/100 g of BF, respectively. The BS showed SP, WSI, and WAI values of 3.5 ± 0.5 (sediment weight g/100 g BS), 1.7 ± 1.6 (weight of the soluble BS g/100 g of BS), and 3.6 ± 0.5 (sediment weight g/weight of BS (dry solid) g), respectively. The SC was higher than SF and was double with respect to starches of other origins. The BPC solubility was 15.5 g protein/100 g BPC (pH 4.5), higher than concentrates of conventional vegetable proteins. The infrared profile showed higher proportions of deployed structures, i.e., β-sheets (22%) and random coils (18.8%), suitable for emulsifying and gelling properties. Results showed bean by-products as an alternative source of ingredients for the food industry.

Keywords: bean; *Phaseolus vulgaris* L.; native starch; protein concentrate

Citation: Alancay, M.M.; Calliope, S.R.; Miranda, R.M.; Samman, N.C. Use of By-Products of Selection Process of Bean (*Phaseolus vulgaris* L.): Extraction of Protein and Starch. *Biol. Life Sci. Forum* **2023**, 25, 10. https://doi.org/10.3390/blsf2023025010

Academic Editors: Claudia M. Haros, Loreto Muñoz and María Dolores Ortolá

Published: 28 September 2023

1. Introduction

Legumes make up a third of the world's largest family of phanerogams, and even though there are 650 genres and numerous species, only around 20 are used for human consumption. These present important nutritional benefits, and they are a source of carbohydrates, fibers, proteins, and minerals, and also contain antinutrients. Currently, there is a great demand for raw materials rich in proteins and carbohydrates for human and animal nutrition, which, in addition, can be obtained at low costs. These demands create the need to develop technologies for the integral use of grains to produce foods and ingredients, including proteins and starch from legumes. The NOA region, including Jujuy province, is an important producer of beans. This activity generates by-products that are mostly not used today. The discarded material is made up mainly of beans with bruises, malformations, and that are broken. It is estimated that between 6 and 8% of the annually harvested beans are lost, which adds up to approximately 7000 tons in the NOA region. In this sense, this study aims to develop technological processes for the integral use of discarded beans through the extraction of protein and starch and their subsequent structural and techno-functional characterization.

2. Materials and Methods

2.1. Raw Materials

Beans (*Phaseolus vulgaris* L.) by-products (BBps) were provided by Cooperativa de Tabacaleros (San Salvador de Jujuy, Argentina). BBps were crushed until their particle size was $\leq$250 μm (Hammer mill KINEMATIC model PX-MFC 90 D, Switzerland), obtaining BBp flour (BF).

2.2. Chemical Composition

Moisture, fat, protein and ash of BF, starch, and protein were determined with Official Methods of Analysis of AOAC [1]. The content of carbohydrates was calculated by difference. The amilose/amilopectin ratio was determined using a method proposed for Jan et al. [2].

2.3. Macromolecules Extraction

2.3.1. Starch Extraction

A BF/water (1:6 p:p) mixture was brought to pH 9.0 with 1N NaOH, stirred for 1 h at 400 rpm at room temperature, then the liquid fraction was separated and left to settle for 16 h to precipitate the bean starch (BS); this was washed three times with distilled water and centrifuged at 2500 rpm. Finally, it was dried at 50 °C.

2.3.2. Protein Extraction

The BF protein, present in the obtained supernatant, was isolated according to the method described by Du et al. [3]. The extraction diagram is shown in Figure 1. The protein extracted from BF was denominated bean protein concentrate (BPC) and was expressed in g protein/100 g of BPC.

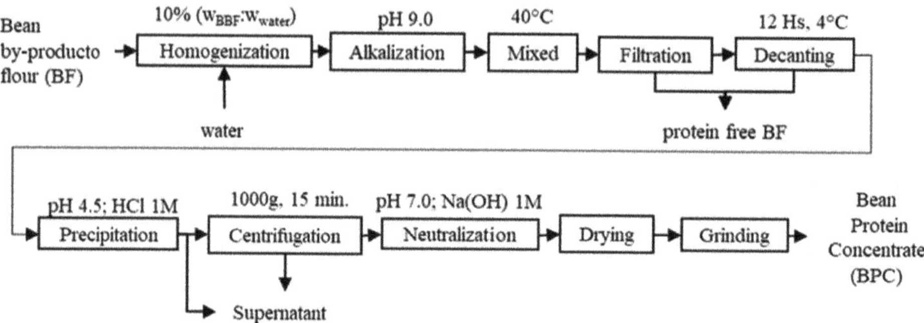

Figure 1. Flow sheet of protein extraction from bean by-product flour (BF).

2.3.3. Extraction Yields

The starch and protein yields were determined using Equation (1).

$$\%\text{Yield} = \frac{W_i}{W_1} \times 100 \tag{1}$$

where W_i is the weight of BS (W_1) or BPC (W_3), and W_1 is BF weight in g.

2.4. Functional Properties

The functional properties were calculated according to Calliope et al. [4].

$$\text{Water absorption index, WAI (g/g)} = \text{weight of sediment/sample weight}$$

$$\text{Water solubility index, WSI (g/g)} = \text{weight of the soluble starch/sample weight} \times 100$$

$$\text{Swelling Power, SP (g/g)} = \text{weight of sediment/(Sample weight-Weight of dissolved solids in supernatant)} \times 100$$

The syneresis in freezing ($-18\ °C$) and refrigeration ($4\ °C$) was determined according to Przetaczek-Rożnowska and Fortuna [5]. The syneresis was measured as % amount of water released.

2.5. Protein Solubility

The solubility of the BPC protein was determined by the method of Du et al. [3] using bovine serum albumin (BSA) to obtain a standard calibration curve (y = 0.0513x + 0.0013; r^2 = 0.9994). The results were expressed as g of soluble protein/100 g BPC.

2.6. Infrared Spectroscopy

The Fourier Transform Infrared (FTIR) spectrum was obtained in the range of 400–4000 cm^{-1} with a resolution of 4 cm^{-1} using an infrared spectrophotometer (Nicolet iS50-thermo Nicolet, Thermo Scientific, United States) with Attenuated Total Reflection (ATR). The infrared spectrum was used to determine the proportions of the secondary structures of BPC.

2.7. Statistical Analysis

The results were subjected to a one-way analysis of variance (ANOVA), and the average was compared using Tukey's test with a confidence level of 95% ($p < 0.05$), using the Infostat statistical software (Version 2015, University of Cordoba, Argentina).

3. Results and Discussion

3.1. Chemical Composition of Bean Flour (BF)

Table 1 shows the BF composition. The moisture was within the average values reported for whole grain flours of different varieties. The amylose content was 28% and was within the range reported by other authors [6]. The crude protein was higher than 19%, which was reported by Sarmento [7] for beans of the same variety. The lipid concentration was less than 2%.

Table 1. Chemical composition of bean flour Alubia variety (BFA), starch (BS), and bean protein concentrate (BPC).

Component (g/100 g)	BFA	BS	BPC
Humidity	11.4 ± 0.1	12.6 ± 0.1	7.7 ± 0.2
Protein	20.4 ± 1.2	0.5 ± 0.05	72.4 ± 0.4
Lipid	1.1 ± 0.01	0.1 ± 0.01	0.6 ± 0.03
Ash	4.7 ± 0.02	0.3 ± 0.01	8.4 ± 0.1
Carbohydrates *	62.5	nd	10.9
Amilose (g/100 g)	-	27.9 ± 0.4	-
Amilopectin (g/100 g)	-	72.1 ± 0.4	-
Phosphorus (mg/100 g BS)	-	95.9 ± 2.6	-
Yield (%)	-	50.3	13.0

* 100 − (humidity + protein + lipid + ash); nd: no determined.

The starch extraction method allowed us to obtain a good quality product due to the low concentration of protein and ash. The BS had a purity value similar to the value reported by Przetaczek-Rożnowska et al. [5].

The extraction yield of the protein was 13%. The BPC had a protein concentration of 72.4%; it was higher than other vegetable sources such as peanuts (42.4%), peas (71.6%), defatted wheat germ, and beer barley by-products (45.7%), among others [8]. The Argentine Food Code established a minimum of 65 g of protein/100 g of sample, to be defined as a "Protein Concentrate".

3.2. Functional Properties of Starch

The determination of functional properties is a method to study the effect of starch process and to characterize the behavior of the BS in food systems. The WAI, WSI, and SP measured in the bean are shown in Table 2. The properties of BS are similar to those of Zaragoza bean starch, except the WSI, whose properties exhibit lower values, probably due to its low amylose content (21.81%) [9]; with respect to potato starch, the BS also presents similar values for these properties, although the WSI was close to the lower limit of the range reported by Calliope et al. [4]. This is probably due to the higher amylose content of BS (27.9%). The amylose content and the length of the amylopectin chains of the starch granules are determining factors of the functional properties [9]. Figure 2 shows the syneresis of BS gels that are stored in frozen and refrigerated conditions. During the first seven days, syneresis was high in both, but significantly higher in the refrigerated gel. Syneresis is considered an undesirable attribute and is associated with an unstable structure. Instability may be due to the rearrangement of the gel matrix or the mechanical damage to the network in a weak gel due to its concentration. In this study, the gel was formed at the same concentration for both treatments, which might attribute the instability to the amylose content that might cause a faster rearrangement in refrigerated condition compared to freezing condition. However, in the freezing condition, the structure formed presented greater stability against freezing–thawing cycles, which might favor a lower and slower syneresis during the measured storage time.

Table 2. Functional properties of bean starch (BS).

Raw Material	WAI **	WSI *	SP *
Bean	3.50 ± 0.54	1.67 ± 1.63	3.56 ± 0.52
Potato [4]	2.31–4.84	1.16–9.54	2.56–4.89
Zaragoza Bean [9]	3.33–4.43	5.70–8.30	3.23–4.43

Values represent mean ± standard deviation. * Expressed in g/100 g BS; ** Values expressed in g/g BS. Source: Calliope et al. [4] and Miranda-Villa et al. [9]. NA: not available.

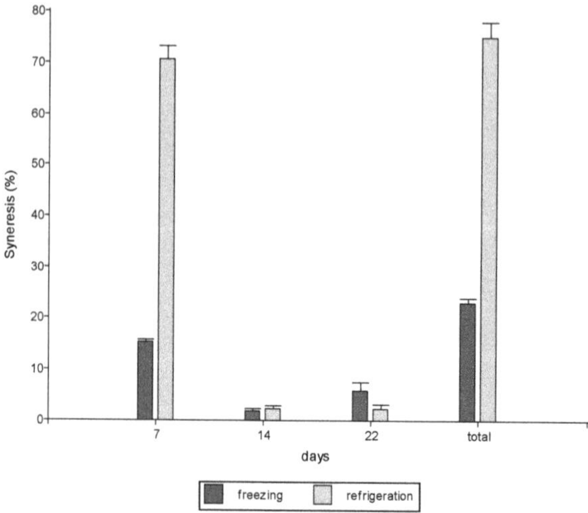

Figure 2. Bean starch syneresis during storage.

3.3. Bean Protein Concentrate (BPC) Solubility

The BPC solubility curve is U-shaped, with a minimum value of approximately 15% at pH 4.5–5.0, corresponding to the isoelectric point (pI); the maximum values were 38.1 and 40.1% at pH 3.0 and 8.0, respectively.

3.4. Profile of FTIR Spectrum of BPC

The infrared spectrum of BPC is shown in Figure 3. The peak present in a region of 3500–3200 cm^{-1} (Amida A) indicates an -NH stretch corresponding to the bending of the vibrational frequency of intra- and intermolecular hydrogen bonding. The peak at 3274.1 cm^{-1} in the BPC spectrum corresponds to a helical formation integrated by hydrogen bonds due to the interaction of functional groups of C=O and N=H. The peak determined at 3076.9 cm^{-1} responds to vibrational stretching of the =C-H bond or vibrations of aromatic C-H bonds from unsaturated hydrocarbons or lipids, respectively. This suggests that the presence of the peak at 3076.9 cm^{-1} could be due to the small trace of lipid present in the BPC. The peaks 2961.1 and 2873.3 cm^{-1}, located between 2990–2850 cm^{-1}, represent the antisymmetric and symmetric methyl (CH_3) and methylene (CH_2) stretching modes that are normally found in aliphatic protein side chains. The three main peaks at 1633, 1530, and 1395 cm^{-1} are assigned to be amide I, amide II, and amide III, respectively. These correspond to C=O stretching, N-H splitting or C-N stretching, and C-N stretching, respectively. The peaks around 1448 and 1239 cm^{-1} are attributed to CH_2 vibrational splitting (scissor type) and C-N stretching, respectively. At 1298.3 cm^{-1}, a peak corresponding to in-plane strain vibrations of -OH is observed. The peak around 1057 cm^{-1} could be assigned to C-O-C antisymmetric stretching variations. Finally, the peak at 509.6 cm^{-1} may correspond to out-of-plane stretching of O-H bonds.

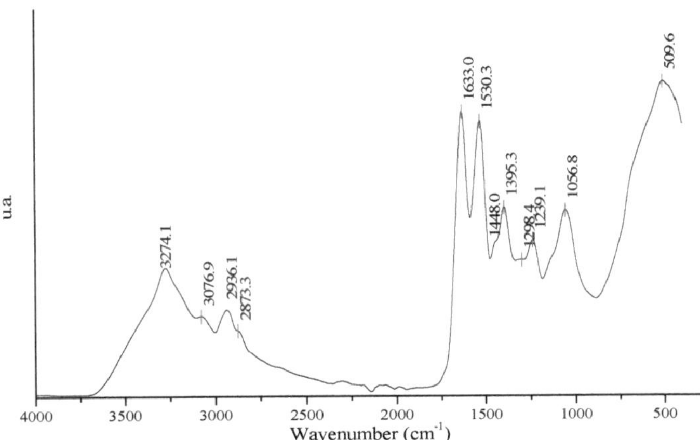

Figure 3. Infrared spectrum of bean protein concentrate (BPC).

3.5. Proportions of Secondary Structures

The deconvolution of the infrared spectrum in the amide I zone allows the determination of the percentage of area under the curve of its populations. Figure 4 shows the populations that comprise the infrared spectrum in the amide I zone, where the identified molecular structures are β-sheet, random coil, α-helix and β-turn, protein aggregates (A1), and amino acid side chain (A2) [10]. Table 3 presents the percentage content of the secondary structures and minor fractions present in the BPC. The secondary structures present in the highest percentage are present in the following order: α-helix, β-sheet, and random coil. The minority structures with lesser extent are for A1, with an exception for A2. The presence of the α-helix (a greater proportion) and A1 structures are associated with an ordered, more compact molecular conformation and with a greater exposure to hydrophilic zones [11]. The β-sheet and random coil structures in BPC impart to it less compact and disordered structure zones together with greater exposure to hydrophobic zones [11]. According to Alancay et al. [11], BPC may have the structural characteristics (most relevant being the presence of β-sheet structure) for its use as an emulsifying and gelling agent in food matrices. The prevalence of the α-helix structure, over the rest of the structures, shows that BPC has greater digestibility compared to the soybean protein isolate [11].

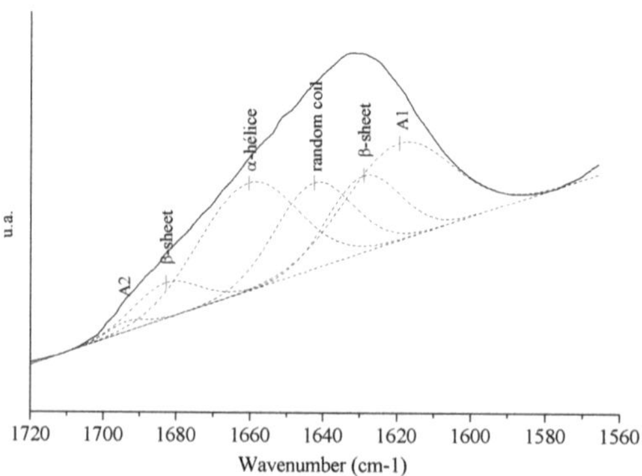

Figure 4. Deconvoluted infrared spectrum for Amide I zone (1750–1550 cm^{-1}) in bean protein concentrate (BPC). The minority fractions A1 and A2 comprise protein aggregates and amino acid side chains, respectively.

Table 3. Area (%) of secondary structures and minor fractions present in the Amide I zone of the bean protein concentrate (BPC).

Area (%)				
Secondary Structures			**Minority Fractions**	
β-sheet (1625–1640 cm^{-1})	Random coil (1680–1690 cm^{-1})	α-helix (1637–1645 cm^{-1})	A2 * (1690–1695 cm^{-1})	A1 * (1610–1625 cm^{-1})
22.0	18.8	29.7	1.0	28.5

* The minority fractions A1 and A2 comprise protein aggregates and amino acid side chains, respectively.

4. Conclusions

The extraction yields of the by-products of bean (*Phaseolus vulgaris*) are presented here, and these by-products may serve as an alternative source of protein and starch. The chemical and physicochemical properties of BS and BPC are within the purity ranges proposed by other authors. The amylose content of the native starch may explain their high syneresis during the first days of storage. Retrogradation is one of the main factors influencing the quality of starch-containing products, usually occurring during storage; SB has a tendency to decline in quality; therefore, its use in food systems that involve refrigeration processes may not be appropriate. On the other hand, the structural properties of the BPC make this material suitable for use as an emulsifying and gelling agent in traditional food systems.

Author Contributions: Conceptualization, M.M.A., S.R.C., R.M.M. and N.C.S.; methodology, M.M.A., S.R.C., R.M.M. and N.C.S.; software, M.M.A., S.R.C., R.M.M. and N.C.S.; validation, M.M.A., S.R.C., R.M.M. and N.C.S.; formal analysis, M.M.A., S.R.C., R.M.M. and N.C.S.; investigation, M.M.A., S.R.C., R.M.M. and N.C.S.; resources, M.M.A., S.R.C., R.M.M. and N.C.S.; data curation, M.M.A., S.R.C., R.M.M. and N.C.S.; writing—original draft preparation, M.M.A., S.R.C., R.M.M. and N.C.S.; writing—review and editing, M.M.A., S.R.C., R.M.M. and N.C.S.; visualization, M.M.A., S.R.C., R.M.M. and N.C.S.; supervision, M.M.A., S.R.C., R.M.M. and N.C.S.; project administration, M.M.A.,

S.R.C., R.M.M. and N.C.S.; funding acquisition, M.M.A., S.R.C., R.M.M. and N.C.S. All authors have read and agreed to the published version of the manuscript.

Funding: This work was funded by Consejo Nacional de Investigaciones Científicas y Técnicas (CONICET-Argentina).

Institutional Review Board Statement: Not applicable.

Informed Consent Statement: Not applicable.

Data Availability Statement: Not applicable.

Acknowledgments: This work was supported by SECTER-CONICET-UNJU (Argentina) and IaValSe-Food-CYTED (119RT0567). Our research group appreciates the collaboration with Mamaní, I M; Cruz, E G; and Gallardo, M.

Conflicts of Interest: The authors declare no conflict of interest.

References

1. AOAC. *Official Methods of Analysis of AOAC International*, 20th ed.; AOAC International: Washington, DC, USA, 2016.
2. Jan, K.N.; Panesar, P.S.; Rana, J.C.; Singh, S. Structural, Thermal and rheological Properties of Starches Isolated from Indian Quinoa Varieties. *Int. J. Biol. Macromol.* **2017**, *102*, 315–322. [CrossRef] [PubMed]
3. Du, M.; Xie, J.; Gong, B.; Xu, X.; Tang, W.; Li, X.; Li, C.; Xie, M. Extraction, Physicochemical Characteristics and Functional Properties of Mung Bean Protein. *Food Hydrocoll.* **2018**, *76*, 131–140. [CrossRef]
4. Calliope, S.; Wagner, J.; Samman, N. Physicochemical and Functional Characterization of Potato Starch (*Solanum Tuberosum Ssp. Andigenum*) from the *Quebrada De Humahuaca*, Argentina. *Starch-Stärke* **2020**, *72*, 1900069. [CrossRef]
5. Przetaczek-Rożnowska, I.; Fortuna, T. Effect of Conditions of Modification on Thermal and Rheological Properties of Phosphorylated Pumpkin Starch. *Int. J. Biol. Macromol.* **2017**, *104*, 339–344. [CrossRef] [PubMed]
6. Betancur-Ancona, D.; López-Luna, J.; Chel-Guerrero, L. Comparison of the Chemical Composition and Functional Properties of Phaseolus Lunatus Prime and Tailing Starches. *Food Chem.* **2003**, *82*, 217–225. [CrossRef]
7. Sarmento, T.R. Impacto Del Procesamiento Sobre La Pared Celular y Las Propiedades Hipoglucémicas y Tecnofuncionales de Leguminosas, Universidad Autónoma de Madrid, Madrid. 2012. Available online: http://hdl.handle.net/10486/12817 (accessed on 31 July 2023).
8. Franca-Oliveira, G.; Fornari, T.; Hernández-Ledesma, B. A Review on the Extraction and Processing of Natural Source-Derived Proteins through Eco-Innovative Approaches. *Processes* **2021**, *9*, 1626. [CrossRef]
9. Miranda-Villa, P.P.; Marrugo-Ligardo, Y.A.; Montero-Castillo, P.M. Caracterización Funcional de Almidón de Frijol Zaragoza (Phaseolus Lunutus) y Cuantificación de su Almidón Resistente. In *Tecnológicas*; Instituto Tecnológico Metropolitano: Medellín, Colombia, 2013; pp. 17–32. Available online: https://www.redalyc.org/pdf/3442/344234332002.pdf (accessed on 15 August 2023).
10. Beck, S.M.; Knoerzer, K.; Arcot, J. Effect of Low Moisture Extrusion on a Pea Protein Isolate's Expansion, Solubility, Molecular Weight Distribution and Secondary Structure as Determined by Fourier Transform Infrared Spectroscopy (FTIR). *J. Food Eng.* **2017**, *214*, 166–174. [CrossRef]
11. Alancay, M.M.; Lobo, M.O.; Samman, N.C. Physicochemical and Structural Characterization of Whey Protein Concentrate–Tomato Pectin Conjugates. *J. Sci. Food Agric.* **2023**, *103*, 5242–5252. [CrossRef] [PubMed]

biology and life sciences forum

Proceeding Paper

An Oxidative Stability Study of Amylose-Hydrolyzed Chia Oil Inclusion Complexes Using the Rancimat Method [†]

Andrea E. Di Marco [1], Vanesa Y. Ixtaina [1,2,*] and Mabel C. Tomás [1]

[1] Centro de Investigación y Desarrollo en Criotecnología de Alimentos (CIDCA), CCT La Plata (CONICET), Facultad de Ciencias Exactas (FCE-UNLP), CICPBA, Calle 47 y 116, La Plata 1900, Argentina; andrea_dimarco@hotmail.com (A.E.D.M.); mabtom@hotmail.com (M.C.T.)

[2] Facultad de Ciencias Agrarias y Forestales (FCAyF-UNLP), Calle 60 y 119, La Plata 1900, Argentina

* Correspondence: vanesaix@hotmail.com

[†] Presented at the V International Conference la ValSe-Food and VIII Symposium Chia-Link, Valencia, Spain, 4–6 October 2023.

Abstract: Chia oil is a source of α-linolenic (omega-3) fatty acid, which is known to promote human health but is highly prone to oxidation. Amylose (a polymer of α-1,4 D-glucose units) can molecularly encapsulate hydrophobic molecules, forming inclusion complexes that could potentially allow the incorporation of sensitive bioactive substances into functional foods. The evaluation of their oxidative stability is relevant to understand their behavior as delivery systems, but monitoring this parameter under real storage conditions requires long periods. In the present work, the oxidative stability of amylose-hydrolyzed chia oil inclusion complexes at 25 °C was estimated from the extrapolation of the exponential dependence of the Rancimat induction times determined at different temperatures (70–98 °C). The complexes were formed with high amylose corn starch and enzymatically hydrolyzed chia oil (10% or 20% hydrolysate/starch), with and without crystallization, using the KOH/HCl method followed by freeze-drying. The spectra of attenuated total reflectance Fourier-transform infrared spectroscopy revealed typical bands that confirmed the effective retention of chia oil fatty acids by the starch structure. The scanning electron micrographs showed that these samples were formed by irregular and porous solid particles. The induction time at 25 °C of crystallized complexes decreased with an increasing hydrolysate content, while the opposite was observed in non-crystallized complexes, as those formed with 20% hydrolysate were the ones that showed the highest stability. Although these findings should be confirmed under real storage conditions, the Rancimat results could be considered as a preliminary quick prediction of the behavior of inclusion complexes as carriers of omega-3 fatty acids.

Keywords: amylose inclusion complex; chia seed oil; omega-3; Rancimat

Citation: Di Marco, A.E.; Ixtaina, V.Y.; Tomás, M.C. An Oxidative Stability Study of Amylose-Hydrolyzed Chia Oil Inclusion Complexes Using the Rancimat Method. *Biol. Life Sci. Forum* **2023**, *25*, 11. https://doi.org/10.3390/blsf2023025011

Academic Editors: Claudia M. Haros, Loreto Muñoz and María Dolores Ortolá

Published: 28 September 2023

1. Introduction

Omega-3 fatty acids are known for their multiple benefits, such as decreasing the risk of cardiovascular and inflammatory diseases and eye health promotion. Chia (*Salvia hispanica* L.) seed oil is a vegetable-rich source of α-linolenic acid (omega-3, ~63%). However, it is highly susceptible to oxidative deterioration, which restricts its incorporation into food products.

During the last few years, there have been carried out several research studies aimed at developing delivery systems capable of protecting sensitive bioactive compounds against deteriorating conditions (T, O_2, pH, light). Among them, amylose inclusion complexes (ICs) have demonstrated suitable functional properties as potential carriers of several molecules. They are formed by amylose, the linear component of starch, which can self-assemble into a single left-handed helix, hosting the gust molecule (ligand) within its inner hydrophobic cavity [1]. In previous work, it was shown that chia oil fatty acids successfully formed inclusion complexes under different times and temperatures of crystallization [2]. Later research demonstrated that the accelerated oxidative stability of these crystallized structures depended on the initial ligand concentration [3], but information about their protective capacity under real storage conditions is still lacking. However, the evaluation of this parameter requires long periods of analysis. In this sense, the present work aimed to study the stability of crystallized and non-crystallized inclusion complexes (10% and 20% w/w ligand/starch) at 25 °C by using the Rancimat method. Moreover, the non-crystallized ICs were characterized by attenuated total reflectance Fourier-transform infrared spectroscopy and scanning electron microscopy. The results of this research would contribute to understanding the behavior of inclusion complexes as potential vehicles of omega-3 from chia oil.

2. Materials and Methods

2.1. Materials

Ingredion Inc. (Westchester, IL, USA) provided high amylose corn starch. Commercial chia oil obtained by cold pressing was purchased from Solazteca SDA S.A. (Buenos Aires, Argentina). The lipase from *Candida rugosa* (type VII, >700 units/mg) was purchased from Sigma-Aldrich (St. Louis, MO, USA). All reagents used were of analytical grade.

2.2. Formation of Inclusion Complexes

Before the complex formation, the chia oil was hydrolyzed by following the procedure previously described [2]. The *Candida rugosa* lipase was incorporated into a chia oil-in-water emulsion (20% w/w), and the reaction was carried out for 5 h under continuous stirring at 37 °C and pH = 7. The hydrolysate obtained was separated by acidification with HCl 2N to pH < 2 and centrifugation (4000× g, 20 min). The non-crystallized inclusion complexes formed with a 10% and 20% w/w hydrolysate/starch (H/S) ratio (NC10 and NC20) were obtained according to the alkaline methodology previously described [3], but without heating the complexes at 90 °C. Additionally, for the oxidative stability test, inclusion complexes crystallized for 2 h at 90 °C with 10% and 20% w/w H/S (C10 and C20) were also formed, according to a previous protocol [3]. The ICs were freeze-dried, ground with a mortar, and sieved.

2.3. Attenuated Total Reflectance Fourier-Transform Infrared Spectroscopy (ATR-FTIR) of Non-Crystallized Complexes

The ATR-FTIR spectra of non-crystallized complexes were obtained using an ATR-FTIR Thermo Nicolet iS10 spectrometer (Thermo Scientific, Waltham, MA, USA). The operating conditions were a 500–4000 cm^{-1} wavenumber range, with 4 cm^{-1} spectral resolution. The results were recorded by co-adding 16 scans and analyzed using OMNIC software (version 8.3, Thermo Scientific, MA, USA).

2.4. Scanning Electron Microscopy (SEM) of Non-Crystallized Complexes

The morphology of the NC10 and NC20 samples was studied using a scanning electron microscope model MA10 (Carl Zeiss SMT Ltd., Cambridge, UK). The powdered samples were adhered to a slide covered with a thin Au/Pd film and examined at 5 kV under high vacuum conditions.

2.5. Oxidative Stability Using the Rancimat Method

The induction times (t_{ind}) of both crystallized and non-crystallized complexes were determined at 70, 80, 90, and 98 °C under 20 L air/h using a Rancimat 743 (Metrohm AG, Herisau, Switzerland) containing 0.5 g of inclusion complexes. Data were analyzed using the 743 Rancimat v1.1 software (Metrohm AG, Herisau, Switzerland). The obtained results from t_{ind} at the different temperatures were fitted to an exponential mathematical model (Equation (1)) using the OriginPro 9.0 software (OriginLab Corporation, Northampton,

MA, USA). To determine the t_{ind} 25 °C, a frequently used room storage temperature, an extrapolation to 25 °C was performed:

$$t_{ind} = A \times e^{(B \times T)} \tag{1}$$

where t_{ind} is the induction time; A and B are the coefficients of the exponential formula; and T is the temperature (°C).

3. Results and Discussion

3.1. Attenuated Total Reflectance Fourier-Transform Infrared Spectroscopy (ATR-FTIR)

The ATR-FTIR spectra of the non-crystallized inclusion complexes formed with 10% and 20% w/w hydrolysate/starch ratios are shown in Figure 1. As can be seen, both samples showed bands at 2852 and 2924 cm^{-1}, which originated from the symmetric and asymmetric vibrations of the -C-H bonds of the CH$_2$ groups of the alkyl chains of fatty acids, respectively [4]. In addition, they also displayed a band at ~1707 cm^{-1}, corresponding to the stretching vibration of the carbonyl (-C=O) of the acid group of fatty acids [4]. The presence of these typical bands in the IC spectra confirms the effective retention of chia fatty acids by both complexes [5], which could be present both in the internal cavity of the amylose helix and/or physically trapped in the interstitial spaces of adjacent helices [6].

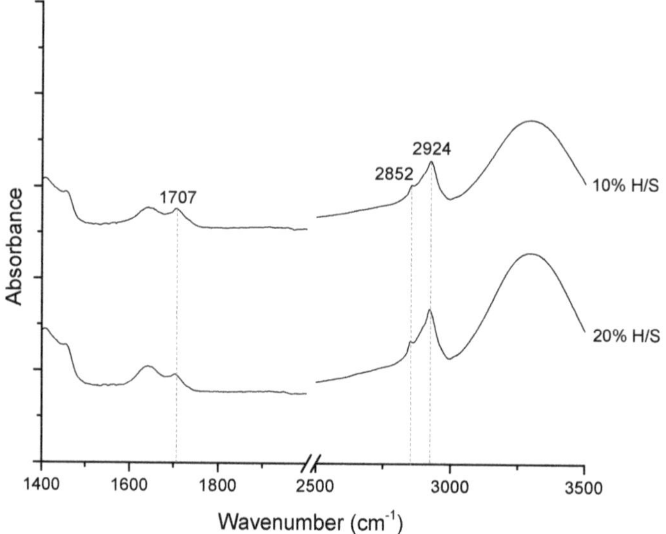

Figure 1. ATR-FTIR spectra of non-crystallized inclusion complexes formed with 10% and 20% w/w hydrolysate/starch. H/S: hydrolysate/starch ratio.

Although the hydrolyzed chia oil used as a ligand had a high percentage of polyunsaturated fatty acids (PUFAs) [3], the characteristic band at 3010 cm^{-1} originating from the =C-H vibration of the *cis* double bonds was not observed in the spectra of the ICs (Figure 1). This may be due to the retention of unsaturated fatty acids into the starch structure, which may shield their IR absorption [7].

3.2. Scanning Electron Microscopy (SEM)

The scanning electron micrographs of the ICs under study are shown in Figure 2. SEM images revealed that they are formed by solid particles of variable sizes (<500 μm). In all cases, an irregular shape and a porous surface (Figure 2) were observed, in agreement with typical micrographs of amylose–fatty acid inclusion complexes formed by the alkaline method and freeze-dried [3,8]. The obtained micrographs are similar to those of the crystallized amylose–chia oil fatty acid ICs [3], and no morphological differences were observed between the non-crystallized complexes formed with 10% or 20% H/S, indicating that neither the omission of the heating at 90 °C during the formation process nor the increase in the H/S ratio modified the morphological properties of the ICs at the microscopic level.

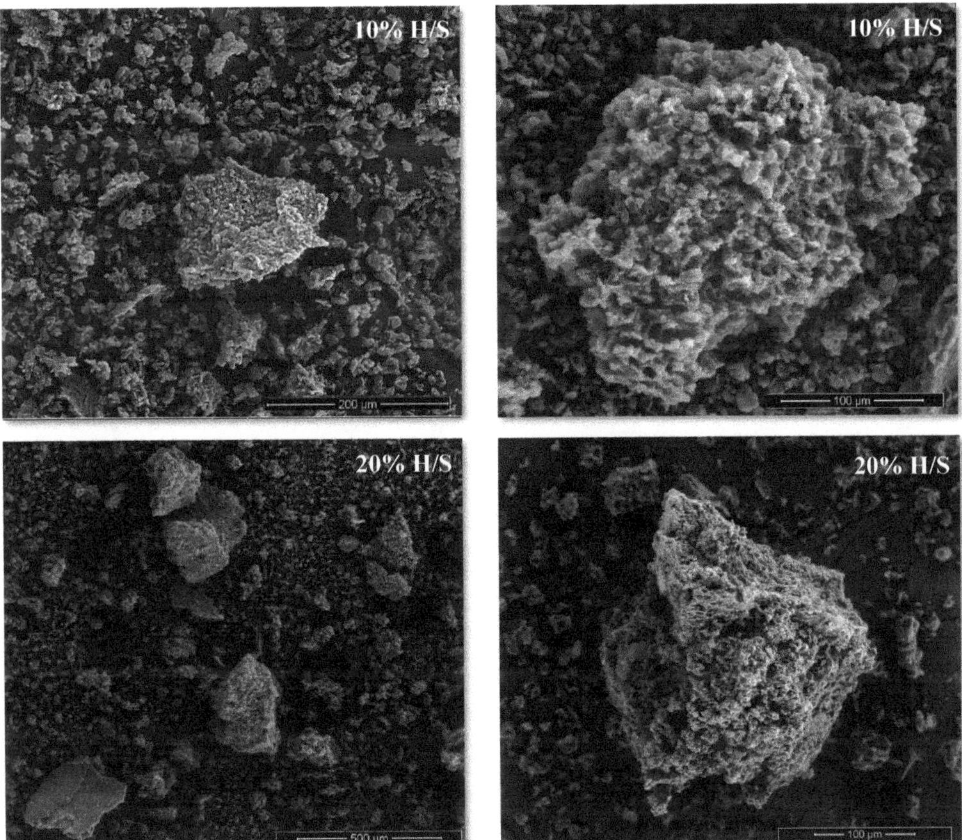

Figure 2. Scanning electron micrographs of non-crystallized inclusion complexes formed with 10% and 20% *w/w* hydrolysate/starch. H/S: hydrolysate/starch ratio.

3.3. Oxidative Stability by Rancimat

The stability of inclusion complexes at 25 °C was estimated modelling the t_{ind} obtained by the accelerated Rancimat method at different temperatures. The results obtained are shown in Table 1. To assess the impact of formation temperature on the oxidative stability of complexes, the analysis also incorporated inclusion complexes (ICs) that were crystallized for 2 h at 90 °C [3]. This inclusion complements the study by enlarging the scope beyond non-crystallized ICs, which constitute the main focus of the present research. In the crystallized ICs, an increase in the H/S ratio from 10% to 20% led to a decrease in the calculated induction time at 25 °C (t_{25}) (Table 1). This finding could be attributed to the higher percentage of PUFAs present in the C20 samples compared to C10 [3], which may be more accessible to oxygen. In contrast, the H/S increase resulted in a higher t_{25} in the non-crystallized ones. This is similar to the previous results of ATR-FTIR, which suggested that the higher initial hydrolysate concentration did not seem to increase the PUFA content of the non-crystallized complexes. The enhanced stability of NC20 in contrast to NC10 may stem from distinct interactions between amylose and fatty acids. Nevertheless, further investigation is required to fully elucidate these differences. Among all the ICs studied, NC20 was the one that showed the highest t_{25}, indicating the most effective protective effect on chia oil fatty acids against oxidation. By not subjecting the formation process to heating at 90 °C, the degradation of polyunsaturated fatty acids (PUFAs) might have been avoided, potentially leading to prolonged stability of the ICs at 25 °C. However, it should be taken into account that the induction time at 25 °C obtained by Rancimat is a theoretical value that serves as a quick estimate of the oxidative stability. Therefore, these results should be validated under real storage conditions.

Table 1. Parameters obtained from the exponential modeling of induction time vs. temperature and induction time extrapolated at 25 °C (t_{25}) of crystallized and non-crystallized inclusion complexes formed with 10% and 20% *w/w* hydrolysate/starch.

Treatment	Hydrolysate/Starch (% *w/w*)	A	B	r^2	t_{25} (h)
Crystallized	10	517.39	−0.0604	0.928	114.3
	20	266.89	−0.0587	0.984	61.5
Non-crystallized	10	82.09	−0.0512	0.934	22.8
	20	1332.56	−0.0750	0.999	204.4

4. Conclusions

According to the results of the present work, chia oil fatty acids were successfully retained in non-crystallized inclusion complexes, as verified by ATR-FTIR. These complexes were formed by irregular and porous solid particles, whose morphological properties did not vary with the different initial hydrolysate concentrations. The non-crystallized inclusion complexes formed with a 20% H/S ratio were those that showed the highest oxidative stability by the Rancimat method. Although further research is still needed, these preliminary results suggest their potential as effective carriers of chia oil essential fatty acids.

Author Contributions: Conceptualization, A.E.D.M., V.Y.I. and M.C.T.; methodology, A.E.D.M. and V.Y.I.; validation, A.E.D.M. and V.Y.I.; formal analysis, A.E.D.M. and V.Y.I.; investigation, A.E.D.M.; resources, V.Y.I. and M.C.T.; writing—original draft preparation, A.E.D.M.; writing—review and editing, V.Y.I.; visualization, A.E.D.M. and V.Y.I.; supervision, V.Y.I. and M.C.T.; project administration, M.C.T. and V.Y.I.; funding acquisition, M.C.T. and V.Y.I. All authors have read and agreed to the published version of the manuscript.

Funding: This research was funded by a grant from la ValSe-Food (119RT0567), the Universidad Nacional de La Plata (UNLP) (11/X907), the Agencia Nacional de Promoción Científica y Tecnológica (PICT 2020-01274) (Argentina), and the Consejo Nacional de Investigaciones Científicas y Técnicas (PIP 2007) (Argentina).

Institutional Review Board Statement: Not applicable.

Informed Consent Statement: Not applicable.

Data Availability Statement: Not applicable.

Acknowledgments: Authors wish to thank Ingredion Inc. for the donation of high amylose corn starch.

Conflicts of Interest: The authors declare no conflict of interest.

References

1. Putseys, J.A.; Lamberts, L.; Delcour, A.J. Amylose-inclusion complexes: Formation, identity and physico-chemical properties. *J. Cereal Sci.* **2010**, *51*, 238–247. [CrossRef]
2. Di Marco, A.E.; Ixtaina, V.Y.; Tomás, M.C. Inclusion complexes of high amylose corn starch with essential fatty acids from chia seed oil as potential delivery systems in food. *Food Hydrocoll.* **2020**, *108*, 106030. [CrossRef]

3. Di Marco, A.E.; Ixtaina, V.Y.; Tomás, M.C. Effect of ligand concentration and ultrasonic treatment on inclusion complexes of high amylose corn starch with chia seed oil fatty acids. *Food Hydrocoll.* **2023**, *136*, 108222. [CrossRef]
4. Guillen, M.D.; Cabo, N. Infrared spectroscopy in the study of edible oils and fats. *J. Sci. Food Agric.* **1997**, *75*, 1–11. [CrossRef]
5. Cao, Z.; Woortman, A.J.; Rudolf, P.; Loos, K. Facile synthesis and structural characterization of amylose–fatty acid inclusion complexes. *Macromol. Biosci.* **2015**, *15*, 691–697. [CrossRef] [PubMed]
6. Guo, Z.; Jia, X.; Miao, S.; Chen, B.; Lu, X.; Zheng, B. Structural and thermal properties of amylose–fatty acid complexes prepared via high hydrostatic pressure. *Food Chem.* **2018**, *264*, 172–179. [CrossRef] [PubMed]
7. Guo, T.; Hou, H.; Liu, Y.; Chen, L.; Zheng, B. In vitro digestibility and structural control of rice starch-unsaturated fatty acid complexes by high-pressure homogenization. *Carbohydr. Polym.* **2021**, *256*, 117607. [CrossRef] [PubMed]
8. Zabar, S.; Lesmes, U.; Katz, I.; Shimoni, E.; Bianco-Peled, H. Structural characterization of amylose-long chain fatty acid complexes produced via the acidification method. *Food Hydrocoll.* **2010**, *24*, 347–357. [CrossRef]

biology and life sciences forum

Proceeding Paper

Characterization of Quinoa Fibre-Rich Fractions Isolated via Wet-Milling and Their Application in Food [†]

Andrea Alonso-Álvarez and Claudia Monika Haros *

Instituto de Agroquímica y Tecnología de Alimentos (IATA), Spanish National Research Council (CSIC), Av. Agustín Escardino 7, Parque Científico, Paterna, 46980 Valencia, Spain; a.alonso@iata.csic.es
* Correspondence: cmharos@iata.csic.es
[†] Presented at the V International Conference la ValSe-Food and VIII Symposium Chia-Link, Valencia, Spain, 4–6 October 2023.

Abstract: Dietary fibre intake has beneficial effects on immunonutritional health and prevents the development of chronic non-communicable diseases such as obesity and diabetes, cardiovascular disease, and cancer. Currently, dietary fibre consumption worldwide is below the WHO recommended daily intake of 25 g. An excellent source of dietary fibre is the fibre-rich fractions of quinoa, which have a high technological potential, nutritional value, and biological activity. This fraction can be isolated via wet-milling, which offers a higher yield and recovery of the main chemical components of cereals/pseudocereals with a higher purity than those obtained via dry-milling. The objective of this work was the isolation of fibre-rich fractions of Royal Bolivian quinoa (white, red, and black) obtained via wet-milling and their characterization as technofunctional ingredients in the formulation of cereal-based food products. The extraction yield of the fibre fraction and its proximal chemical composition were determined, in addition to phytic acid content; minerals such as calcium, iron, and zinc; and technofunctional properties (particle size distribution, water and oil holding capacity, and swelling capacity). All fibre fractions isolated via wet-milling could be used as food ingredients. In particular, the fibre-rich fraction of black quinoa contains the highest amount of insoluble fibre. However, from a technological point of view, red quinoa fibre could be the most suitable for inclusion in the formulation of food matrices due to its high water and oil retention capacity, as well as its swelling capacity. The incorporation of a low proportion of quinoa dietary fibre (5–10%) allows increasing the nutritional profile of cereal-based food products.

Keywords: dietary fibre-rich fraction; quinoa fractionation; wet-milling

Citation: Alonso-Álvarez, A.; Haros, C.M. Characterization of Quinoa Fibre-Rich Fractions Isolated via Wet-Milling and Their Application in Food. *Biol. Life Sci. Forum* **2023**, *25*, 12. https://doi.org/10.3390/blsf2023025012

Academic Editors: Loreto Muñoz and María Dolores Ortolá

Published: 28 September 2023

1. Introduction

The consumption of dietary fibre has shown notable health benefits in terms of immunonutritional well-being and in the prevention of chronic conditions including obesity, type 2 diabetes, cardiovascular ailments, and cancer [1,2]. However, the current global intake of dietary fibre falls below the recommended daily amount of 25 g set by the WHO/FAO [3].

A remarkable reservoir of dietary fibre resides within the fibre-rich fractions of quinoa, which not only possesses high technological potential but also boasts considerable nutritional value and biological activity. These fibre-rich fractions can be efficiently extracted through the wet-milling process of quinoa. This method yields greater quantities and enhanced recovery of vital chemical components found in cereals and pseudocereals (as depicted in Figure 1), while maintaining a higher level of purity compared to dry-milling techniques.

The primary aim of our study was to isolate fibre-rich fractions from different varieties of Bolivian quinoa—white, red, and black—utilizing wet-milling techniques. A comprehensive characterization was subsequently carried out, evaluating their potential as technofunctional ingredients in the development of cereal-based food products, including applications such as fresh pasta.

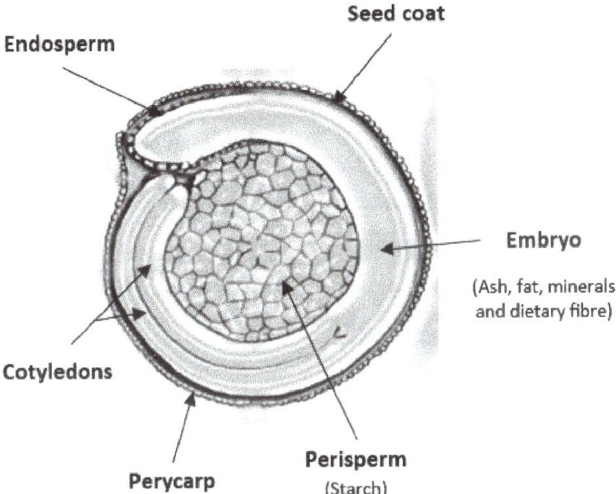

Figure 1. Longitudinal section of quinoa seed adapted from Prego et al. [4].

2. Materials and Methods

2.1. Materials

Royal quinoa grains (white, red, and black) produced in Bolivia were obtained from Organic Quinoa Real® (Figure 2).

Figure 2. Black, red, and white Royal quinoa grains.

2.2. Methods

Wet-Milling

Quinoa seeds were steeped in a sodium bisulfite solution adjusted with lactic acid at pH 5.0, in a Biostat® B plus fermenter from Sartorius BBI Systems. Wet-milling was performed according to the methodology recommended by Ballester-Sánchez et al. [5], with some modifications (Figure 3).

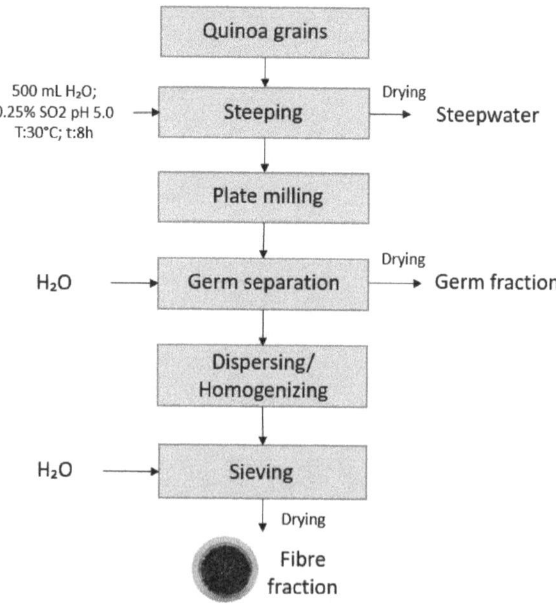

Figure 3. Flow diagram of Royal quinoa fibre isolation via wet-milling, adapted from Ballester-Sánchez et al. [5].

2.3. Chemical Composition

Moisture was determined via gravimetry, starch via the enzymatic/spectrophotometric micromethod [6], protein via combustion Dumas (conversion factor N 5.7) [7], lipids via Randall [8], ash via ignition [9,10], and soluble dietary fibre (SDF) and insoluble dietary fibre (IDF) via the enzymatic/gravimetric method [11,12]. Minerals such as calcium, iron, and zinc and total phosphorous were determined via atomic absorption spectrophotometry [13].

2.4. Fresh Pasta Production

Laboratory-scale tagliatelle were made using the Nina electric pasta machine from Springlane® (Düsseldorf, Germany). All formulations included wheat flour mixed with water and egg yolk. However, four different formulations were prepared: a control made with 100% wheat flour, and three types in which 6% of the wheat flour was replaced by the fibre fractions isolated via wet-milling of white, red, and black quinoa.

3. Results

Extraction yield of the quinoa fibre fraction obtained via wet-milling was between 14 and 21% in dry matter. The proximate composition (protein, lipid, starch, ash, and total dietary fibre (TDF) content) is shown in Figure 4. The fibre fractions show significant differences ($p < 0.05$) with respect to the TDF, with black quinoa having the highest proportion, followed by red quinoa. However, no significant differences were observed in the percentages of SDF between the different fractions (Figure 5). In terms of their technofunctional properties, the red quinoa fibre presented between 40 and 60% more water and oil retention capacity, as well as more swelling power than the rest of the fibres studied.

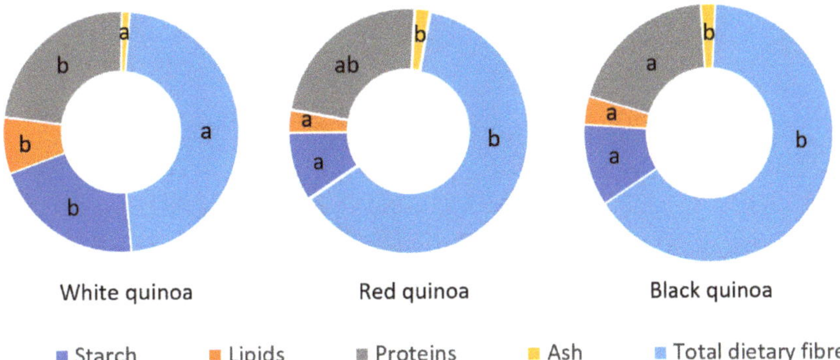

Figure 4. Proximate composition of the fibre-rich fractions of white, red, and black quinoa. Mean ± SD (*n* = 3), expressed in % dry matter. The portion of each parameter with the same letter do not show significant differences at 95% confidence level.

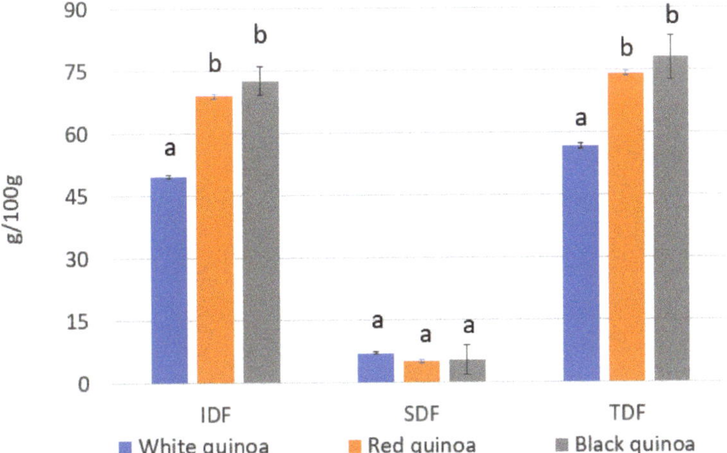

Figure 5. IDF, SDF, and TDF content of white, red, and black Royal quinoa fibre fractions. Mean ± SD (*n* = 3), expressed in % dry matter. The bars of each parameter with the same letter do not show significant differences at 95% confidence level. IDF: insoluble dietary fibre; SDF: soluble dietary fibre; TDF: total dietary fibre.

In general, the quinoa fibre showed high concentrations of minerals Ca, Fe, Zn, and P (Figure 6). No significant differences were found in the phosphorus of the different dietary fibre fractions of the differently coloured quinoas. However, the high phosphorous could indicate a high phytate content, meaning mineral absorption may be compromised due to the formation of insoluble compounds with phytates.

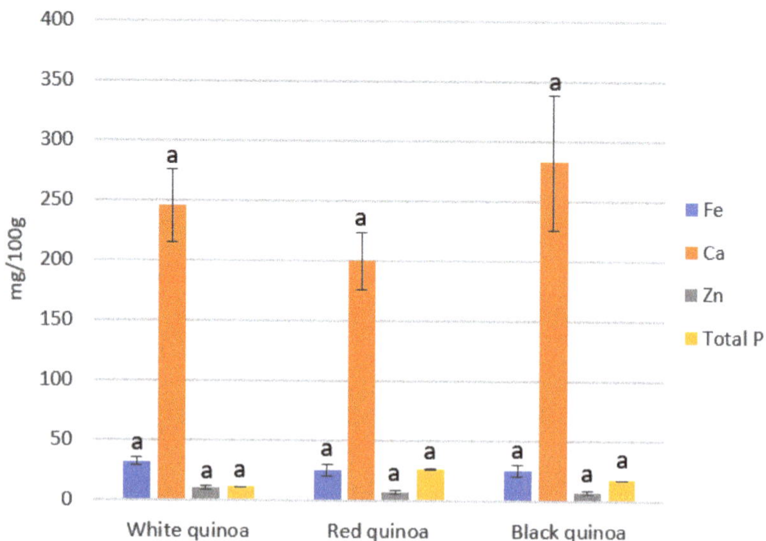

Figure 6. Mineral (Ca, Fe, and Zn) content and total phosphorous in the dietary fibre fraction of Royal quinoa obtained via wet-milling. Mean ± SD (*n* = 3), expressed as mg/100 g dry matter. The bars of each parameter with the same letter do not show significant differences at 95% confidence level.

The intake of 100 g of pasta fortified with a 5–10% fraction of dietary fibre sourced from white, red, or black quinoa fibre can contribute approximately 35% of an adult's daily fibre requirement (Figures 7 and 8).

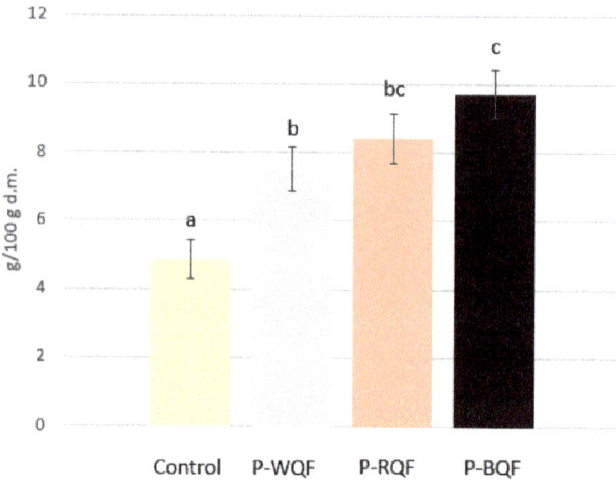

Figure 7. Dietary fibre content in pasta with fibre addition of white, red, and black quinoa. Mean ± SD (*n* = 3), expressed as g/100 g in dry matter (d.m.). The bars with the same letter do not show significant differences at 95% confidence level. Control: pasta control; P-WQF: pasta with white quinoa fibre; P-WRF: pasta with red quinoa fibre; P-BQF: pasta with black quinoa fibre.

Figure 8. Detailed picture of red quinoa fibre pasta formulation.

4. Conclusions

All fibre fractions obtained through the wet-milling process have the potential to serve as valuable food ingredients. Notably, the fibre-rich fraction extracted from black quinoa boasts the highest content of insoluble fibre. However, considering technological aspects, the fibre derived from red quinoa emerges as an optimal candidate for incorporation into food formulations. This is attributed to its capacity for water and oil retention, alongside its ability to swell.

Incorporating a modest proportion of dietary fibre sourced from quinoa can significantly enhance the nutritional value of cereal-based food products.

Author Contributions: Conceptualization, C.M.H.; Funding acquisition, C.M.H.; Investigation, A.A.-Á. and C.M.H.; Methodology, A.A.-Á. and C.M.H.; Project administration, C.M.H.; Supervision, C.M.H.; Writing—original draft, A.A.-Á. and C.M.H.; Writing—review and editing, C.M.H. All authors have read and agreed to the published version of the manuscript.

Funding: This work was supported by grants Ia ValSe-Food-CYTED (Iberoamerican Valuable Seeds—119RT0567) and Food4ImNut (PID2019-107650RB-C21) of the Ministry of Science and Innovation (MICINN-Spain).

Institutional Review Board Statement: Not applicable.

Informed Consent Statement: Not applicable.

Data Availability Statement: The data presented in this study are available on request from the corresponding author. The data are not publicly available due to their privacy.

Acknowledgments: The contract of Andrea Alonso-Álvarez from INVESTIGO Program within the framework of the Recovery, Transformation and Resilience Plan—Government of Spain, in the Generalitat Valenciana, is gratefully acknowledged. The authors sincerely thank Ana Belén Romero Alonso for her important collaboration in generating some of the results presented in this publication.

Conflicts of Interest: The authors declare no conflict of interest.

References

1. Slavin, J. Fiber and Prebiotics: Mechanisms and Health Benefits. *Nutrients* **2013**, *5*, 1417–1435. [CrossRef] [PubMed]
2. Nohra, E.; Bochicchio, G.V. Management of the Gastrointestinal Tract and Nutrition in the Geriatric Surgical Patient. *Surg. Clin.* **2015**, *95*, 85–101. [CrossRef] [PubMed]
3. Jones, J.M. CODEX-aligned dietary fiber definitions help to bridge the 'fiber gap'. *Nutr. J.* **2014**, *13*, 34. [CrossRef] [PubMed]
4. Prego, I.; Maldonado, S.; Otegui, M. Seed Structure and Localization of Reserves in *Chenopodium quinoa*. *Ann. Bot.* **1998**, *82*, 481–488. [CrossRef]
5. Ballester-Sánchez, J.; Gil, J.V.; Fernández-Espinar, M.T.; Haros, C.M. Quinoa wet-milling: Effect of steeping conditions on starch recovery and quality. *Food Hydrocoll.* **2019**, *89*, 837–843. [CrossRef]
6. AOAC Official Method 996.11, Starch (Total) in Cereal Products. Amyloglucosidase-α-Amylase Method First Action 1996, AOAC-AACC Method. Available online: http://www.aoacofficialmethod.org/index.php?main_page=product_info&products_id=1546 (accessed on 15 July 2023).

7.   *ISO/TS 16634-2:2016*; Food Products—Determination of the Total Nitrogen Content by Combustion According to the Dumas Principle and Calculation of the Crude Protein Content—Part 2: Cereals, Pulses and Milled Cereal Products. ISO: Geneva, Switzerland, 2016.

8.   Cereals and Grains Association. 30-25.01 Crude Fat in Wheat, Corn, and Soy Flour, Feeds, and Mixed Feeds. In *AACC Approved Methods of Analysis*, 11th ed.; Cereals and Grains Association: St. Paul, MN, USA, 2000.

9.   *ISO 2171:1980*; Cereals, Pulses and Derived Products—Determination of Ash. ISO: Geneva, Switzerland, 1980.

10.   *ICC 104/1*; Determination of Ash in Cereals and Cereal Products. ICC: Vienna, Austria, 1990.

11.   AOAC Official Method 991.43 Total, Soluble, and Insoluble Dietary Fibre in Foods. 1996. Available online: https://acnfp.food.gov.uk/sites/default/files/mnt/drupal_data/sources/files/multimedia/pdfs/annexg.pdf (accessed on 6 August 2023).

12.   Cereals and Grains Association. 32-07.01 Soluble, Insoluble, and Total Dietary Fiber in Foods and Food Products. In *AACC Approved Methods of Analysis*, 11th ed.; Cereals and Grains Association: St. Paul, MN, USA, 2000.

13.   Miranda-Ramos, K.; Millán-Linares, M.C.; Haros, C.M. Effect of Chia as Breadmaking Ingredient on Nutritional Quality, Mineral Availability, and Glycemic Index of Bread. *Foods* **2020**, *9*, 663. [CrossRef] [PubMed]

biology and life sciences forum

Proceeding Paper

Global Trends in the Worldwide Expansion of Quinoa Cultivation [†]

Didier Bazile [1,2]

[1] CIRAD, UMR SENS, 34398 Montpellier, France; didier.bazile@cirad.fr

[2] UMR SENS, CIRAD, IRD, University Paul Valery Montpellier 3, University Montpellier, 34090 Montpellier, France

[†] Presented as Plenary Conference at the V International Conference la ValSe-Food and VIII Symposium Chia-Link, Valencia, Spain, 4–6 October 2023.

Abstract: For centuries, quinoa cultivation was centered only in the Andean countries, and recently it has spread to all regions of the world. Although the number of exporting countries has increased, Bolivia and Peru remain the world's leading producers and exporters. Today, more than 125 countries are experimenting with or cultivating quinoa. The expansion of the crop has only been possible due to the genetic diversity of seeds maintained by generations of farmers in the Andes. As access to quinoa genetic resources in Andean countries remains limited, this implies that the development of new varieties relies on a narrow genetic base relative to the theoretical potential of the species. The use of improved varieties has increased, especially with the emergence of new countries sourcing seed from commercial varieties to start cultivation. To cope with the increasing effects of climate change, it is essential to increase the resilience of crop by taking advantage of their genetic diversity. The current global crisis can only be overcome in the North or in the South by establishing new partnerships for access to genetic resources and the fair and equitable sharing of the benefits of their use. In the last 30 years, quinoa from the Andean countries gained a position in global markets and improved the quality of life of producers. However, at the end of 2015, producer prices collapsed. Quinoa development is dynamic, and now Andean producers face different scenarios with new competitors and new concerns. Being aware of the new reality is essential to face the new challenges responsibly. Analysis at different scales is fundamental, as is promoting local diversity and cooperating towards innovative production systems and inclusive processes that benefit everyone.

Keywords: Andes; biodiversity; cultivation expansion; markets; producers; quinoa

Citation: Bazile, D. Global Trends in the Worldwide Expansion of Quinoa Cultivation. *Biol. Life Sci. Forum* **2023**, 25, 13. https://doi.org/10.3390/blsf2023025013

Academic Editors: Claudia M. Haros, Loreto Muñoz and María Dolores Ortolá

Published: 28 September 2023

1. Introduction

Biodiversity conservation today is a key global concern of the international community with the last (IPBES, IPCC, FAO) global assessments in 2019. This loss of biodiversity places our agriculture and our food at risk. In Latin America, the Andean altiplano is one of the centers of origin or "hot-spots" of the world's biodiversity. For thousands of years, the populations have interacted with the agroecosystems. Quinoa crop has evolved from a complex process of biological, geographical, climatic, social and cultural interactions that have determined its current high genetic diversity.

The Pluri-national State of Bolivia has requested the FAO to declare 2013 "International Year of Quinoa (IYQ)". By resolution 66/221 of 22 December 2011, the United Nations General Assembly declared 2013 the International Year of Quinoa (IYQ) and the secretariat was assigned to FAO-RLC (Santiago de Chile). The IYQ aimed to draw global attention to the role of quinoa's biodiversity and nutritional value in food security and poverty eradication in order to achieve the Millennium Development Goals.

Quinoa (*Chenopodium quinoa* Willd.), known as a neglected and underutilized species, was considered a major crop used by the Pre-Colombian cultures in Latin-America for centuries. As a consequence of the invasion and the conquest by the Spanish, cultivation and consumption of this crop were suppressed and thereafter only continued on a minor or local scale. Quinoa has been grown in the Andes for over 7000 years. After centuries of neglect, the potential of quinoa was only rediscovered during the second half of the twentieth century [1]. Following the International Year of Quinoa (IYQ) in 2013, the case of quinoa was highlighted with the potential to rapidly change its status from a minor to a major crop in the world agriculture, on basis of the role that quinoa's biodiversity and its high nutritional value can play in providing global food security [2].

But the question at the heart of global expansion is the following: "Can the nutritional richness of quinoa serve the global food security?". Compared to the major cereals for agriculture and world food (wheat, corn, rice), quinoa has a much higher protein content (from 14 to 19%). But above all, it presents a good balance between all the essential amino acids, with contents above the FAO recommendations for each of them. Much emphasis has been be placed on the quality of quinoa's proteins for its promotion and worldwide recognition, but its nutritional value is more global [3].

The balanced structure of essential amino acids is one of the main characteristics of quinoa, but not the only one. Quinoa grains contain a very high proportion of polyunsaturated fatty acids (especially omegas 3, 6, 9), essential for human growth and development (brain, muscles, retina), as well as very high levels of Vitamin E or tocopherols as powerful antioxidants (α, β, γ and δ) and tocotrienols (α, β, γ and δ) with known biological activity, essential for reproduction and growth of mammals including humans. α-tocopherol in particular is a powerful antioxidant that prevents the oxidation of plant lipids, especially seeds (it also aids in the fight against cholesterol). It protects the body from cell damage and helps maintain a healthy immune system to protect against chronic diseases, such as heart disease and cancer. Quinoa is also very rich in many minerals, and it contains as much fiber as whole grains, useful to moderate the glycemic index of the meal. Vitamin B1 allows the production of energy from carbohydrates (sugars). Vitamin B9 or folic acid (also called folates) is essential for cell renewal and for the development of the fetus during pregnancy.

2. Quinoa Genetic Diversity

The answer to healthy eating lies in the diversity cultivated and maintained by small-scale farmers in the Andes. Quinoa was domesticated near Lake Titicaca between Peru and Bolivia. Generations of farmers have been involved in quinoa selection, which explains the high levels of genetic diversity found today. Quinoa diversity, at a continental scale, has been associated with five main ecotypes: Highlands or Altiplano (Peru and Bolivia), Inter-Andean valleys (Colombia, Ecuador and Peru), Salt flats (Bolivia, Chile and Argentina), Tropical Yungas (Bolivia) and Coastal/Lowlands (Chile). Each of these ecotypes is associated with sub-centres of diversity that comes from the surroundings of Lake Titicaca. And each one corresponds to specific conditions of altitude, latitude and soils and climatic conditions [4–6]. Due to its extraordinary genetic diversity, the crop is highly adaptable to different agroecological conditions and tolerant to frost, drought and salinity [3]. The differences are very pronounced between ecotypes. Each of these ecotypes is associated to specific conditions of altitude, latitude and is adapted to specific soils and climatic conditions. Considering its origin in central and southern Chile, the sea level ecotype appears as the most adapted to temperate and Mediterranean environments. With high attention to the Chilean germplasm, the number of quinoa-producing countries has risen rapidly from 8 (in the 80') until 125 today [7,8].

3. Phases of Quinoa Spreading

From at least 5000 years ago until the beginning of the 1980s, Quinoa has been specific to the Andean countries in South America. But since the other countries understood the potential and benefits of quinoa, the amount of experimentation conducted did not stop growing.

During the 1980s, the multiplication and the spread of experimental stations were directly linked to major international initiatives for research. Research partnerships have often facilitated the exchange of germplasm and have had a powerful impact on this development by strengthened collaborations. However, partnerships between research institutions for germplasm exchanges need to consider legal and ethical aspects related to the access to genetic resources for experimentation and fair commercial development. The first introduction of quinoa in Europe began in 1978, with germplasms coming from Chile (Universidad de Concepción), and then selected and tested in Cambridge, England. From Cambridge, quinoa was distributed to Denmark, The Netherlands and other countries.

The most important project during the 1990s that explains today's quinoa worldwide expansion is the project with the Danish International Development Agency (DANIDA) and the International Potato Center (CIP) in Peru. There were fields trials in new countries, and most of them were involved in the global European and American Test of Quinoa, organized by the FAO (Food and Agriculture Organization of the United Nations). The aim of this project was to create a state-of-the-art quinoa based on multiple experimentations at a global level.

The spread of worldwide quinoa was caused by strong relationships between institutions that shared their genetic material. FAO played a key role on this issue during the International Year of Quinoa.

During the past thirty years, quinoa was tested in all the continents, and nowadays, quinoa is cultivated in more than 125 countries [9]. Quinoa globalization entails challenges to the countries of origin, and these are important to consider for future development. Understanding this reality is fundamental to face the challenges of conserving local biodiversity, developing and promoting new varieties, and cooperating on plant genetic resources exchanges with inclusive processes towards fair benefits with Andean countries [10].

4. The New Quinoa Producers

Who are the new producers of quinoa? Most of them (>90%) are introducing quinoa as a new crop for diversifying their cropping system. And they cultivate less than 2 ha of quinoa and less than 30% of their superficies. Less than 25% of the new superficies are cultivated with irrigation, especially around the Mediterranean sea. We note more organic production in the Andes and conventional production outside. For introducing quinoa, considering the difficulties of accessing quinoa genetic diversity from Andean countries, commercial varieties are mainly used, and new country producers are registering new varieties with Plant Variety Rights (UPOV System) [11].

With the global demand, increasing production in the countries of origin for international market put at risk the agriculture–livestock integration in Andean agroecosystems. But at the same time, from the 1990s to 2008, farmer incomes have been multiplied by 6!; the attractiveness of quinoa has generated new social conflicts over access and land sharing between Andean communities.

There is a strong interest in quinoa in the Mediterranean and in developing countries. It has two main objectives: to help fight malnutrition and to help reduce poverty, because quinoa grains have a high nutritional value and it is a rustic crop than can grow from the sea level to over 4000 m of altitude.

5. Discussion

Today, seed legislations at a global level limits the access to genetic quinoa resources for testing the crop in new environments. Legal restrictions are important for international regulations on seeds and plant genetic resources so that only a very small part of the available genetic diversity is used for the adaptation of quinoa to new environments. Only 3 to 12 local varieties (with high intrinsic diversity) are usually tested simultaneously, and only 1 to 3 certified commercial varieties (Puno, Titicaca) are widely distributed.

First, *a question of fair access to genetic resources*. Andean countries hold the largest germplasm collections. But many countries have established collections prior to the signing of the Convention on Biological Diversity which specifies that states are sovereign over their genetic material. For quinoa, however, there is no single existing legal framework providing a comprehensive coverage of all the issues related to the genetic resources and their sustainable management. The Convention on Biological Biodiversity (CBD) does regulate bilateral access and benefit sharing, but this is difficult to apply to quinoa as a crop is now planted internationally, not restricted to the Andean region, and this has been the case for decades. This now means that these countries may well develop new varieties from this germplasm without having to refer directly to the country of origin. But one main problem today is the lack of transparency in PGR flows at a global level. Often, the main objective of the trials is not so clear for some countries. There is a need for better defining the objectives of the variety trials [12]. An objective of adaptation from an existing commercial variety differs completely from a strategy of plant breeding for developing new adapted cultivars. Germplasm and experimental design for the trials may be adapted to these two distinct objectives. The Andean peasant varieties of quinoa are heterogeneous and well adapted to extreme climate and soil conditions due to a very high intra-varietal genetic diversity.

Second, *a question of access to technologies.* Some knowledge and technologies are already available, but there is an imbalance technology access between the North and the South. Key factors for the expansion of the quinoa and future improvement and breeding are the following items: using molecular markers (SSR linkage map, marker-assisted selection), improving feature selection based on genes of interest, PVB/PPB methods, adaptation to climate change and salinity using variability.

Third, *a question of sustainable development.* How can we consider sustainable agriculture development for food security and nutrition with NUS? Quinoa is a good example of an adaptive crop for many environments that can help to restore agricultural systems in marginal and degraded areas, considering its tolerance to salinity and drought. But how can we improve resource efficiency of production, and also of natural resources (soil, biodiversity, water, etc.)? To achieve sustainable development goals, we need to think about agricultural systems resilience against risk, against variability and against uncertainty. At the same time, social aspects also need to be considered, like land access and sharing, seed exchanges, cooperatives and farmers' organization for transforming quinoa and assessing to markets, etc. In addition, new country producers have difficulties introducing quinoa grains in their national markets considering post-harvesting operations. Therefore, training both farmers and agricultural advisors is essential for developing a complete quinoa value chain. Elaborating and testing local dishes with quinoa from the beginning of the programs could be key for the appropriation of the quinoa as a food crop for the population and for its integration into their diet.

6. Conclusions

Drought resistance, salinity tolerance and exceptional nutritional value are some of the advantages of quinoa facing the effects of climate change in agriculture. But access to genetic resources is necessary to allow the adaptation of an exotic species in new environments. Research plays a central role in the development of quinoa through international collaborations. It is necessary, however, to be patient before expecting commercial production. Adoption of quinoa by local populations is essential for producing it in a sustainable way.

Funding: This research received no external funding.

Institutional Review Board Statement: Not applicable.

Informed Consent Statement: Not applicable.

Data Availability Statement: Not applicable.

Acknowledgments: I want to thank here all the members of the Global Collaborative Network on Quinoa (GCN-Quinoa.org) who are active actors in developing quinoa at a global level. Thanks also to FAO for providing assistance to countries interested in testing quinoa as an alternative crop for fighting with global changes and for increasing food security. As a quinoa international focal point for FAO, I am helping these countries to offer technical assistance with the presence of good previous thinking regarding their objectives.

Conflicts of Interest: The authors declare no conflict of interest.

References

1. Bazile, D. *La quínoa. Los Desafíos de una Conquista*; LOM Ediciones: Santiago, Chile, 2016.
2. FAO & CIRAD. *State of the Art Report on Quinoa Around the World in 2013*; Bazile, D., Bertero, H.D., Nieto, C., Eds.; Food and Agriculture Organization of the United Nations (FAO): Santiago, Chile, 2015.
3. Ruiz, K.B.; Biondi, S.; Oses, R.; Acuña-Rodríguez, I.S.; Antognoni, F.; Martinez-Mosqueira, E.A.; Coulibaly, A.; Canahua-Murillo, A.; Pinto, M.; Zurita, A.; et al. Quinoa biodiversity and sustainability for food security under climate change. A review. *Agron. Sustain. Dev.* **2014**, *34*, 349–359. [CrossRef]
4. Fuentes, F.F.; Martinez, E.A.; Hinrichsen, P.V.; Jellen, E.A.; Maughan, P.J. Assessment of genetic diversity patterns in Chilean quinoa (Chenopodium quinoa Willd.) germplasm using multiplex fluorescent microsatellite markers. *Conserv. Genet.* **2009**, *10*, 369–377. [CrossRef]
5. Fuentes, F.F.; Bazile, D.; Bhargava, A.; Martinez, E.A. Implications of farmers' seed exchanges for on-farm conservation of quinoa, as revealed by its genetic diversity in Chile. *J. Agric. Sci.* **2012**, *150*, 702–716. [CrossRef]
6. Bazile, D.; Fuentes, F.; Mujica, A. Historical perspectives and domestication. In *Quinoa: Botany, Production and Uses*; Bhargava, A., Srivastava, S., Eds.; CABI Publisher: Wallingford, UK, 2013; pp. 16–35.
7. Bertero, H.D.; De la Vega, A.J.; Correa, G.; Jacobsen, S.E.; Mujica, A. Genotype and genotype-by-environment interaction effects for grain yield and grain size of quinoa (Chenopodium quinoa Willd.) as revealed by pattern analysis of international multi-environment trials. *Field Crops Res.* **2004**, *89*, 299–318. [CrossRef]
8. Christensen, S.A.; Pratt, D.B.; Pratt, C.; Nelson, P.T.; Stevens, M.R.; Jellen, E.N.; Coleman, C.E.; Fairbanks, D.J.; Bonifacio, A.; Maughan, P.J. Assessment of genetic diversity in the USDA and CIP-FAO international nursery collections of quinoa (*Chenopodium quinoa* Willd.) using microsatellite markers. *Plant Genet. Resour.* **2007**, *5*, 82–95. [CrossRef]
9. Alandia, G.; Rodriguez, J.P.; Jacobsen, S.E.; Bazile, D.; Condori, B. Global expansion of quinoa and challenges for the Andean region. *Glob. Food Sec.* **2020**, *26*, 100429. [CrossRef]
10. Andreotti, F.; Bazile, D.; Biaggi, M.C.; Callo-Concha, D.; Jacquet, J.; Jemal, O.M.; King, O.I.; Mbosso, C.; Padulosi, S.; Speelman, E.N.; et al. When neglected species gain global interest: Lessons learned from quinoa's boom and bust for teff and minor millet. *Glob. Food Sec.* **2022**, *32*, 100613. [CrossRef]
11. Bazile, D.; Jacobsen, S.E.; Verniau, A. The global expansion of quinoa: Trends and limits. *Front. Plant Sci.* **2016**, *7*, 622. [CrossRef] [PubMed]
12. Bazile, D.; Pulvento, C.; Verniau, A.; Al-Nusairi, M.; Ba, D.; Breidy, J.; Hassan, L.; Maarouf, I.M.; Mambetov, O.; Otambekova, M.; et al. Worldwide evaluations of quinoa: Preliminary results from post international year of quinoa FAO projects in nine countries. *Front. Plant Sci.* **2016**, *7*, 850. [CrossRef] [PubMed]

biology and life sciences forum

Proceeding Paper

Development of a Grain-and-Legume-Based Snack with Amaranth, Quinoa and Chia Seeds †

Silvia Farah, Jannika Bailey, Pablo Mezzatesta and Emilia Raimondo *

Faculty of Nutrition Sciences, Juan Agustín Maza University, Mendoza M5519, Argentina;
farahsilvia1@hotmail.com (S.F.); jannikabailey@gmail.com (J.B.); pmezzatesta@umaza.edu.ar (P.M.)
* Correspondence: emilia.raimondo@gmail.com
† Presented at the V International Conference la ValSe-Food and VIII Symposium Chia-Link, Valencia, Spain, 4–6 October 2023.

Abstract: Globally, 1 in 2 people want to consume more plant-based foods, so developing nutritionally balanced foods for this area is a response to a growing need of the population. The objective was to develop snack alternatives with amaranth, quinoa and chia seeds, supplemented with protein. For this purpose, three mixtures were prepared, the first of rice flour with lentil flour (M1), another of rice flour with pea flour (M2) and another of wheat flour with chickpea flour (M3). The same amount of amaranth, quinoa and chia seeds were added to all three. The same proportion of the rest of the ingredients was used, varying only the seasonings. The cooking time was 10 min at 180 °C in all cases. The preparations were carried out in triplicate and analyses in duplicate. Once prepared, protein, total fat, ash, moisture and fiber were determined using official analytical techniques, carbohydrates using difference, energy value using calculations and fatty acid profile using gas chromatography. For statistical analysis, the Tukey test was applied. The highest protein value was 15.84 ± 0.34 g% (M1). The highest lipid value was 21.44 ± 0.13 g% (M3), providing omega 9, 3 and 6 fatty acids and, to a much lesser extent, saturated fatty acid. All options have a good contribution of dietary fiber: 7.39 ± 0.14 g% (M1); 7.71 ± 0.09 g% (M2) and 7.55 ± 0.23 g% (M3). The acceptability was above 94% in all cases. It was possible to formulate healthy snacks suitable for vegetarians.

Keywords: amaranth; chia; nutritional profile; quinoa; snack; vegetable protein

Citation: Farah, S.; Bailey, J.; Mezzatesta, P.; Raimondo, E. Development of a Grain-and-Legume-Based Snack with Amaranth, Quinoa and Chia Seeds. *Biol. Life Sci. Forum* **2023**, *25*, 16. https://doi.org/10.3390/blsf2023025016

Academic Editors: Claudia M. Haros, Loreto Muñoz and María Dolores Ortolá

Published: 8 October 2023

1. Introduction

Consumption habits are changing worldwide, with one in two people preferring to consume plant-based proteins. These changes are due to respect for animal suffering, religious beliefs, care for the environment or fashion. In Argentina, vegetarians account for 12% of the population, while worldwide they represent 7%. Flexitarians consume animal proteins sporadically and represent 24% of the world's population [1].

Within the category of vegetarians, there are different subgroups, depending on the type of food they eat [1]:

Vegetarian (strict): They do not consume any type of meat or animal by-products, such as dairy products, eggs and honey.

Vegan: Like the previous description, they do not consume any food of animal origin, but they do not accept any product that is the result of animal suffering, such as wool, leather, silk, etc., nor do they accept products that are tested on animals.

Api-ovo-lacto-vegetarian: In addition to plant-based foods, they include honey, eggs and dairy products.

Api-lacto-vegetarian: Includes vegetables, dairy and honey (excluding eggs).

Api-vegetarian: Includes vegetables and honey (excludes eggs and dairy).

Api-ovo-vegetarian: Includes vegetables, egg and honey (excludes eggs and dairy).

Ovo-vegetarian: Includes plant-based foods and eggs.

Lacto-vegetarian: Includes plant-based foods and dairy.

Raw vegetarian "raw vegan": Consume everything raw, including fruits, vegetables, nuts, seeds, sprouted legumes, cereals, sprouts, etc.

Raw vegan "raw vegan": Same as above, but with an added ethical commitment. The critical nutrients for strict vegans are: high quality proteins, calcium, zinc and vitamin B12 [2,3]. Therefore, developing nutritionally balanced foods for this population segment, and for the rest of the consumers who, without being vegetarians, want to reduce their intake of meat proteins, is a response to a growing need in the population.

Amaranth seed was legislated in article 660 of the Argentinean Food Code (C.A.A.) which states the following: "The name Amaranth means the healthy, clean and well preserved seeds of the following species of this pseudo-cereal: *Amaranthus cruentus* L., *Amaranthus hipochondriacus* L., *Amaranthus caudatus* L. and *Amaranthus mantegazzianus Passer*. The protein content should not be less than 12.5%, the moisture content should not be more than 12.0%, the ash content should be less than 3.5%, the starch content should not be less than 60%. Amaranth grains, corresponding to the above mentioned species, shall be white, pale amber, yellow or very pale brown, opaque or translucent" [4].

Chia seeds were legislated in Article 918 (C.A.A.), which reads as follows: "The name chia seeds means healthy, clean and well-preserved seeds of *Salvia hispanica* L. They shall comply with the following specifications: chia seeds, which correspond to the species mentioned, shall be dark brown in colour, very small in size and of good fluidity. The aroma shall be mild, pleasant and characteristic of the seed. Determination of humidity, at 100–105 °C: maximum 7%. Fat: minimum 33%. They must not contain more than 0.5% of damaged seeds. They shall be free from live insects. They shall contain not more than 1% foreign matter, of which not more than 0.25% shall be mineral material and not more than 0.10% dead insects, fragments or remains of insects and/or other impurities of animal origin" [4].

Quinoa seed was legislated in Article 682 (CAA), which states the following: "The name quinoa or quinoa means sound, clean and well-preserved seeds of the genus *Chenopodium quinoa* Willd. They shall comply with the following specifications: Total protein on dry basis: minimum 10%. Moisture: maximum 13.5%. Ash on dry basis: maximum 3.5%. The quinoa or quinoa seeds to be industrialized must be subjected to a process that ensures the elimination of saponins and the bioavailability of amino acids. For seeds marketed in the absence of the customer, the legend "Wash until removal of foam. Not suitable for raw consumption, cook before consumption" [4].

The food called "snack biscuits" is legislated in article 760 quarter CAA [4].

For all the above, the objective was to develop snack alternatives with amaranth, quinoa and chia seeds, supplemented with protein.

2. Materials and Methods

In a first stage, the optimal ratio of the mixture of rice flour with lentil flour (M1), rice flour with pea flour (M2) and wheat flour with chickpea flour (M3) was calculated. This was carried out by chemical computation in order to obtain the best biological value of the mixtures, which are also technologically suitable. Once this calculation had been carried out, the snack biscuits were prepared.

2.1. Ways of Preparation

At the Applied Nutrition Research Laboratory of Juan Agustín Maza University, Mendoza, Argentina, the snack biscuits were prepared on a pilot scale. For this purpose, the mixtures were prepared in triplicate. The ingredients of the rice–lentil mixture (M1), rice–pea mixture (M2) and wheat with chickpea (M3) can be seen in Table 1. In all cases, water was used to integrate the ingredients. In the three variants, all components were integrated into a homogeneous dough, which was rolled out, cut into different shapes in order to identify them, and then cooked. The three mixtures were cooked for 10 min at

180 °C, then cooled and packaged at room temperature in 40 micron polypropylene bags. Centesimal composition analyses were performed in duplicate on the packaged snacks.

Table 1. Ingredients of the three mixtures.

Rice–Lentil Mixture (M1)	Rice–Pea Mixture (M2)	Wheat with Chickpea (M3)
33.86% rice flour	33.50% rice flour	26.47% wheat flour
18.14% lentil flour	18.29% pea flour	26.47% chickpea flour
11% extra virgin olive oil	11.07% extra virgin olive oil	10.61% extra virgin olive oil
2.86% amaranth seeds	2.86% amaranth seeds	2.86% amaranth seeds
2.86% quinoa seeds	2.86% quinoa seeds	2.86% quinoa seeds
1.43% chia seeds	1.43% chia seeds	1.43% chia seeds
0.36% white pepper	0.29% oregano	0.35% garlic
0.36% salt	0.29% salt	0.32% dried parsley
0.29% sugar	0.29% garlic	0.25% salt
0.29% paprika	0.29% dried parsley	0.25% sugar
28.57% water	0.29% sugar	0.20% turmeric
	28.57% water	27.93% water

2.2. Laboratory Analysis

The analyzes were carried out at time zero, at 30 days and at 60 days of packaging. The following methods were used to determine the nutritional composition of the snacks M1, M2 and M3 [5]:

Humidity: Method of AOAC 950.46 B. Indirect method by drying in an oven at 100–105 °C, until constant weight is achieved.

Total fat: Direct method by extraction with ethyl ether (crude fat), Soxhlet gravimetric method (A.O.A.C. 960.39, 1990) was used.

Fibers: Acid alkaline attack.

Crude protein: Kjeldahl method, (A.O.A.C. 928.08, 1990), determining nitrogen, using 6.25 as a protein conversion factor.

Ashes: Direct Method (A.O.A.C. 923.03, 1990): by incineration in muffle (at 500 ± 10 °C), until constant ash weight. *Sodium* by flame photometry on ash dilution.

Carbohydrates: determined by difference, using the following formula:

100 − (weight in grams [protein + fat + water + ash + fibers]), in 100 g of food.

Energy value: by calculation

(kcal) = (protein × 4) + (carbohydrates × 4) + (fat × 9). The conversion is 2000 kcal = 8400 kJ.

Fatty Acid profile: using gas chromatography according to the official methods of the IOC (International Olive Oil Council), ISO 5508-1990.

Sensory analysis

For the sensory evaluation, an acceptability test was conducted with 100 untrained judges, using a 5-point scale ranging from "I like it very much" to "I dislike it very much". The evaluators observed the color, smell, texture and appearance of the snacks.

2.3. Statistical Analysis

Only descriptive statistics reporting means and standard deviation were performed, because it is not the intention to compare the three snacks with each other, given that they are three different options. Only the Tukey test for moisture content was applied.

3. Results

For the three snack variants, nutrients are reported on a dry basis. The results at time zero have been placed in Table 2.

Table 2. Nutritional profile of the three snacks.

	Rice–Lentil Mixture (M1)	Rice–Pea Mixture (M2)	Wheat with Chickpea (M3)
Humidity [g]	23.74 ± 0.76	22.88 ± 0.88	24.88 ± 0.56
Proteins [g]	15.84 ± 0.34	14.72 ± 0.35	15.05 ± 0.24
Total fat [g]	18.66 ± 0.13	20.11 ± 0.17	21.44 ± 0.13
Saturated fat [g]	3.01 ± 0.02	3.15 ± 0.05	3.21 ± 0.07
Polyunsaturated fat [g]	3.21 ± 0.03	3.90 ± 0.04	4.64 ± 0.03
Monounsaturated fat [g]	12.43 ± 0.08	13.16 ± 0.09	13.60 ± 0.08
Carbohydrates without fiber [g]	55.89 ± 0.17	55.30 ± 0.25	53.90 ± 0.13
Total sugars [g]	4.90 ± 0.04	7.28 ± 0.04	7.51 ± 0.05
Added sugars [g]	0.43 ± 0.01	0.43 ± 0.01	0.43 ± 0.01
Fibers [g]	7.39 ± 0.14	7.71 ± 0.09	7.55 ± 0.23
Ashes [g]	2.22 ± 0.03	2.16 ± 0.06	2.05 ± 0.07
Sodium [mg]	219 ± 3	182 ± 4	203 ± 2
Energy value [kcal/kJ]	347/1457 kJ	356/1493	352/1479

3.1. Moisture

M3 had a higher moisture content, perhaps because it was formulated with wheat, whose starch has a higher percentage of amylopectin. When applying the T Student test for independent samples, a statistically significant difference can be observed in: moisture $p = 0.003$.

3.2. Proteins

The highest protein contribution was presented by M1, due to the protein contribution of lentils.

3.3. Lipid Profile

M3 has the highest fat content. In the three mixtures, the content of monounsaturated fatty acids provided by the extra virgin olive oil is very important.

3.4. Carbohydrate Profile, without Considering Fiber

In carbohydrates, there are statistically significant differences between M3 and the other two mixtures. M1 has the lowest contribution of total sugars, due to the ingredients used. The added sugars are the same in the three mixtures since the same proportion of sugar was used in all cases.

3.5. Fiber Content

All options have a good contribution of dietary fiber.

3.6. Energy Value

The energy value of snacks are lower than those on the market whose energy intake varies from 430 kcal to 550 kcal per 100 g.

3.7. Sensory Evaluation

The acceptability was above 94% in all cases. Dividing the responses into "I like it very much" and "I like it", the results are shown in Figure 1.

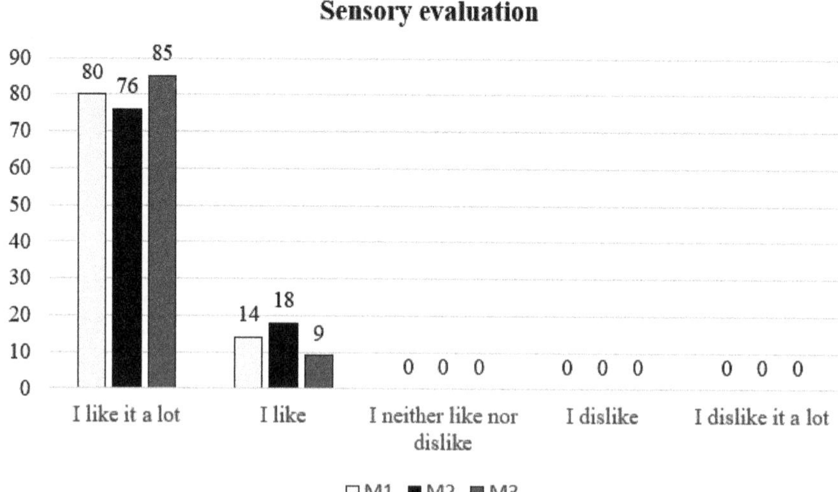

Figure 1. Sensory evaluation of the snacks.

4. Discussion

By mixing cereals and pulses, the biological value of the three snacks increased, and the addition of seeds also increased the total protein content. Both wheat and rice have low biological value proteins, but, when these were supplemented with legumes, such as lentils, peas and chickpeas, they are transformed into proteins of high biological value, ideal for the target group of vegetarians [6,7].

An analysis of the lipid content shows a high content of omega-9 fatty acids from extra virgin olive oil, which is beneficial to health, in addition to the increase in polyunsaturated fatty acids provided by the seeds [8].

The decrease in total carbohydrate content, with a consequent increase in fiber, provided by the seeds, contributes to increased satiety [9,10].

The sodium content can be reduced by not adding salt at the time of preparation and replacing it with seasonings.

5. Conclusions

We were able to formulate three snack biscuits, intended especially for vegetarians, but that can also be consumed by the entire population.

They have an excellent protein content of high biological value, resulting from the combination of cereals and legumes. Rice and wheat are deficient in lysine, which is provided by legumes, thus increasing their biological value. The addition of ancestral seeds improves the nutritional profile.

This increases the supply of healthier foods and enhances the value of ancestral seeds.

Author Contributions: Conceptualization, P.M. and E.R.; methodology, S.F. and J.B.; software, P.M.; validation, S.F., J.B. and P.M.; formal analysis, E.R.; investigation, J.B. and S.F.; resources, E.R.; data curation, P.M.; writing—original draft preparation, E.R.; writing—review and editing, J.B.; visualization, S.F.; supervision, P.M.; project administration, E.R.; funding acquisition, E.R. All authors have read and agreed to the published version of the manuscript.

Funding: This research was funded by Universidad Juan Agustin Maza, Argentina and La ValSe-Food-CYTED (119RT0567).

Institutional Review Board Statement: Not applicable.

Informed Consent Statement: Not applicable.

Data Availability Statement: The article contains all the trial data.

Acknowledgments: This work was supported by a grant from La ValSe-Food-CYTED (119RT0567) and Universidad Juan Agustin Maza, Argentina.

Conflicts of Interest: The authors declare no conflict of interest.

References

1. Alvarez, A.; Brett, C.; Ganduglia, M.; Raspini, M.; Rey, L.; Rodriguez García, V.; Schuldberg, J.; Tassiello, E. Revisión Bibliográfica: Alimentación Vegetariana en la Infancia y Adolescencia. *Diaeta* **2021**, *39*, 59–71. Available online: http://www.scielo.org.ar/scielo.php?script=sci_arttext&pid=S1852-73372021000100059&lng=es&tlng=es (accessed on 13 June 2023).
2. Bakaloudi, D.R.; Halloran, A.; Rippin, H.L.; Oikonomidou, A.C.; Dardavesis, T.I.; Williams, J.; Wickramasinghe, K.; Breda, J.; Chourdakis, M. Intake and adequacy of the vegan diet. A systematic review of the evidence. *Clin. Nutr.* **2021**, *40*, 3503–3521. [CrossRef] [PubMed]
3. Gibson, R.S.; Heath, A.L.; Szymlek-Gay, E.A. Is iron and zinc nutrition a concern for vegetarian infants and young children in industrialized countries? *Am. J. Clin. Nutr.* **2014**, *100* (Suppl. 1), 459S–468S. [CrossRef] [PubMed]
4. ANMAT. Código Alimentario Argentino. 2023. Available online: https://www.argentina.gob.ar/anmat/codigoalimentario (accessed on 10 May 2023).
5. AOAC. Official Methods of Analysis. 1990. Available online: https://archive.org/stream/gov.law.aoac.methods.1.1990/aoac.methods.1.1990_djvu.txt (accessed on 17 May 2023).
6. Carbajal Azcona, A. Manual de Nutrición y Dietética. In *Chapter 5: Proteínas*; Departamento de Nutrición, Facultad de Farmacia, Universidad Complutense de Madrid: Madrid, Spain, 2013; 367p.
7. Cervilla, N.; Mufari, J.; Calandri, E.; Guzman, C. Determinación del contenido de aminoácidos en harina de quinoa de origen argentino. In *Evaluación de su Calidad Proteica*; Actualización en Nutrición; 13; 2; Sociedad Argentina de Nutrición: Buenos Aires, Argentina, 2012; pp. 107–113.
8. Hernández Pérez, T.; Valverde, M.E.; Orona Tamayo, D.; Paredes Lopez, O. Chia (Salvia hispanica): Nutraceutical Properties and Therapeutic Applications. *Proceedings* **2020**, *53*, 17. [CrossRef]
9. Myrisis, G.; Aja, S.; Haros, C.M. Substitution of Critical Ingredients of Cookie Products to Increase Nutritional Value. *Biol. Life Sci. Forum* **2022**, *17*, 15. [CrossRef]
10. Pino Villalón, J.; Rojas Muñoz, M.; Orellana Saez, B.; Torres Mejías, J. Fortificación con fibra dietética como estrategia para aumentar la saciedad: Ensayo aleatorizado doble ciego controlado. Rev. Española Nutr. *Humana Dietética* **2020**, *24*, 336–344. [CrossRef]

Proceeding Paper

Use of Phenolic Extract from Peanut Skin as a Natural Antioxidant in Chia Oil-Based Mayonnaise [†]

Romina Mariana Bodoira [1,*], Andrea Carolina Rodríguez-Ruiz [2], Damián Modesto Maestri [3], Pablo Daniel Ribotta [1], Alexis Rafael Velez [2] and Marcela Lilian Martinez [3]

[1] Instituto de Ciencia y Tecnología de los Alimentos Córdoba, Consejo Nacional de Investigaciones Científicas y Técnicas, Facultad de Ciencias Exactas Físicas y Naturales, Universidad Nacional de Córdoba (ICYTAC—CONICET—FCEFyN—UNC), Av. Juan Filloy S/N, Córdoba 5000, Argentina; pribotta@agro.unc.edu.ar

[2] Instituto de Investigación y Desarrollo en Ingeniería de Procesos y Química Aplicada, Consejo Nacional de Investigaciones Científicas y Técnicas, Facultad de Ciencias Exactas Físicas y Naturales, Universidad Nacional de Córdoba (IPQA—CONICET—FCEFyN—UNC), Av. Vélez Sarsfield 1611, Córdoba 5000, Argentina; carodruiz6@gmail.com (A.C.R.-R.); avelez@unc.edu.ar (A.R.V.)

[3] Instituto Multidisciplinario de Biología Vegetal, Consejo Nacional de Investigaciones Científicas y Técnicas, Facultad de Ciencias Exactas Físicas y Naturales, Universidad Nacional de Córdoba (IMBIV—CONICET—FCEFyN—UNC), Av. Vélez Sarsfield 299, Córdoba 5000, Argentina; dmaestri@unc.edu.ar (D.M.M.); marcela.martinez@unc.edu.ar (M.L.M.)

*   Correspondence: romina.bodoira@mi.unc.edu.ar

[†] Presented at the V International Conference la ValSe-Food and VIII Symposium Chia-Link, Valencia, Spain, 4–6 October 2023.

Citation: Bodoira, R.M.; Rodríguez-Ruiz, A.C.; Maestri, D.M.; Ribotta, P.D.; Velez, A.R.; Martinez, M.L. Use of Phenolic Extract from Peanut Skin as a Natural Antioxidant in Chia Oil-Based Mayonnaise. *Biol. Life Sci. Forum* **2023**, *25*, 17. https://doi.org/10.3390/blsf2023025017

Academic Editors: Claudia M. Haros, Loreto Muñoz and María Dolores Ortolá

Published: 11 October 2023

Abstract: Currently, the antioxidants (AOs) used in foods are mainly synthetic, often questioned on health grounds. So, the need for innocuous natural AOs has increased in last years. The peanut skin (PS) is an industrial by-product of low added value, but rich in bioactive phenolic compounds. In this study, the antioxidant capacity of a PS extract (PSE) was examined in a chia oil-based mayonnaise stored during six months (25 °C). The mayonnaise was made using chia oil (68% w/w) added with PSE (2 mg/g fat) and without any AO (control). For the storage test, 30 g were placed in 100 mL amber bottles and at 2, 4 and 6 months the oily phase was extracted (chloroform: methanol). Peroxide index (PI), acidity index (AI), K_{232}, K_{270}, p-Anisidine (pAnV) and TOTOX values were measured. Moreover, the presence of 2,4 heptadienal and 3,5-octadiene-2-one was analyzed by static headspace GC-MS. At the end of the assay, PI, AI, K_{232}, K_{270}, and pAnV for control and PSE mayonnaises were 74.7 and 13.4 meq O_2/kg oil; 2.4 and 2.0 mg KOH/kg oil, 10 and 3.55, 1.34 and 0.64, 3.7 and 0.98, respectively. The TOTOX value of the control was approximately six times higher than PSE mayonnaise. 2,4 Heptadienal and 3,5-octadiene-2-one were not detected at initial time but in the control treatment at the end reached 3.75 and 2.15 µg/g, respectively. Differently, in PSE mayonnaise, 3,5-octadiene-2-one was undetected and 2,4 Heptadienal was 0.83 µg/g. In conclusion, PSE represents a potential natural AO to preserve the oxidative stability of chia oil-based mayonnaise.

Keywords: natural antioxidant; mayonnaise; byproduct valorization; phenolic compounds; oxidative stability; storage assay

1. Introduction

Mayonnaise, a widely used semi-solid dressing, is an O/W emulsion with a low pH and 65–85% (w/w) of fat. It is formulated by mixing vegetable oil (dispersed phase), water and vinegar (continuous phase), egg yolk (emulsifier), salt and sugar. Mayonnaise is susceptible to spoilage because to autooxidation, but this mechanism is more complex than oxidation in bulk oils due the great oil area exposed to an aqueous phase that turns interfacial oxidation relevant. Moreover, many factors such as oil droplet size, processing conditions, pH, surface charge and area, affect lipid oxidation in emulsion systems [1].

The use of antioxidants (AOs) is very common in the food industry. Synthetic AOs such as butylated hydroxytoluene (BHT), butylated hydroxyl anisole (BHA), tertiary butylhydroquinone (TBHQ) are widely used to delay or prevent lipid oxidation. Some of them, however, are being questioned because both in vitro and in vivo studies have demonstrated possible toxicity. Moreover, the unavoidable AOs release to the environment may cause potential risks to both human and environment health [2]. Some natural polyphenolic compounds have the ability of reducing the rate of lipid oxidation; so, they represent an alternative to synthetic AOs.

The peanut skin (PS), an industrial waste of peanut blanching operations, is an abundant source of phenolic compounds, mostly monomeric and condensed flavonoids. Among the attractive bioactivities of PS phenolic extracts, the antioxidant capacity is probably the most studied [3]. Much of food applications of PS phenolic-based extracts (PSE) have been made on meat products [4], as chemical preservatives that improve oxidative and microbiological spoilage. Additional applications could include their use as AOs in emulsions or others multiphase systems. In this line, Verstraeten et al. [5] found that procyanidin dimers and trimers isolated from PS can protect the integrity of lipid bilayers. According to these, recently, PSE showed good antioxidant efficacy when tested in a highly unsaturated lipid system such as a chia O/W model acid emulsion [6].

As previously stated, the objective of the present study was to evaluate PSE antioxidant efficacy during chia oil-based mayonnaise storage.

2. Materials and Methods

2.1. Chia Oil and Peanut Skin Extract (PSE) Extraction

Chia oil was obtained by cold pressing using a Komet screw press (Komet CA 59G; IBG Monforts, Monchengladbach, Germany). The PS was acquired from peanut Runner-type using a typical industrial blanching process (90 °C, 10 min). Skins were cleaned, milled and sieved to obtain uniform particle sizes (0.5 mm). Phenolic compounds were extracted from PS by means of subcritical fluid extraction following conditions reported previously (220 °C, flow 7 g/min and pressure 7 MPa) [7]. Distilled water and ethanol (40:60, v/v) were used as solvents. After the extraction process, the liquid extract was centrifuged (15 min at 9.000 rpm) and then concentrated under vacuum (40 °C).

Finally, the remaining residue was lyophilized and the obtained dry extract (PSE) was stored (-20 °C) in amber glass bottles under nitrogen. This extract has already been characterized for its antioxidant properties and phenolic composition [6,7].

2.2. Preparation of Chia Oil Based Mayonnaise

Mayonnaise batches of 250 g were prepared using a universal mixer. The ingredients and their proportions were those used by Alizadeth et al. [8] with some modifications. Each batch contained: 170.30 g (68.12% w/w) of chia oil, 35.75 g (14.30% w/w) of water, 2.03 g (0.81% w/w) of vinegar, 22.85 g (9.14% w/w) egg yolk, 2.50 g (1% w/w) lemon juice, 12.50 g (5% w/w) sugar, 3.80 g (1.52% w/w) salt and antimicrobial agents: sodium benzoate and potassium sorbate 0.15 g (0.06% w/w) each one. At first, egg yolk powder and dry materials were mixed manually with water and vinegar. Then, oil was added slowly to the aqueous phase and mixed until a homogeneous emulsion was obtained. For PSE treatment preparation, the PSE was dissolved in 1.5 mL of water:ethanol (40:60 v/v) and subsequently added to the aqueous phase, giving a final PSE concentration of 2 mg/g total fat (oil + egg yolk lipids). This concentration was chosen based on previous studies in an O/W acid emulsion model [7]. Finally, the control treatment was prepared using AOs free chia oil following the same procedure.

2.3. Physical Properties of Chia Oil Based Mayonnaise

The physical properties of the mayonnaise such as color, pH, water activity (a_w) and particle size distribution were measured following the methodologies previously reported [9].

2.3.1. Color Measurements

Mayonnaise samples were measured by CIE L* a* b* scale using a portable colorimeter with a D65 illuminant at a 10° observation (spectrophotometer Konica-Minolta, Tokyo, Japan). Black and white calibrations were necessary before the measurements. Then, an amount of each mayonnaise was poured into the measurement plate and the L* (lightness 0 to 100), a* (red/green coordinate) and b* (yellow/blue coordinate) parameters were recorded. The whiteness index (WI = L* − 3b*) and yellowness index (YI = 142.86 b*/L*) were calculated.

2.3.2. pH and a_w Measurements

pH was measured with a pH meter (Hanna Instruments S.A., Buenos Aires, Argentina) and the a_w by means of a Decagon AquaLab (208 series 3, Decagon Devices Inc., Pullman, WA, USA, EUA water activity meter) equipment at 25 °C.

2.3.3. Particle Size Measurements

Forty mg of mayonnaise sample was mixed with 150 mL of a 0.1% sodium dodecyl sulfate (SDS) solution and stirred until it was completely dispersed. The size distribution of oil droplets was determined by laser diffraction with the Horiba analyzer at 25 °C. The refractive indexes used were 1.47 and 1.33 of oil and water, respectively. The droplet size as the volume-weighted mean diameter ($d_{4.3}$) and the span number were reported.

2.3.4. Texture

Texture profile analysis (TPA) was carried out using a TA-TX2i Texture Analyzer (Universal Testing Machine Model 3342, Instron, EUA). The sample were scoped into the 25 mm cylindrical probe. The test was performed with one cycle at a speed of 1.0 mm s^{-1} to compress the sample to 80% of its original height. The firmness and adhesiveness of the samples are reported.

2.4. Chia Oil Based Mayonnaise Storage Assay

After preparation (Section 2.2), 30 g of the mayonnaise were put into 100 mL brown bottles with film caps that allows some gas exchange. These samples were stored at ambient temperature (25 °C) in a chamber and sampling in triplicate was performed after 2, 4 and 6 months for further analyses. Moreover, for the determination of volatile compounds (VC) an aliquot of mayonnaise (10 g), from both treatment (PSE and control), was placed into a 40 mL headspace vial light cutlery, with silicon septum and sealed cap, and stored for 6 months under the same conditions.

2.4.1. Volatile Compounds by Static Headspace Gas Chromatography

The samples were prepared by adding 3 mL of a sodium chloride saturated solution to break the emulsion [8]. After shaking, 20 µL of internal standard (cyclohexanol 3.8 mg/mL methanol) was added, then the vials were equilibrated (40 min at 50 °C).

Volatile compounds (VC) were sampled from the headspace using a syringe (Hamilton. Model: 81,456 volume: 2.5 mL. Reno, NV, USA) and immediately injected into a PerkinElmer Clarus 600T chromatograph-mass spectrometer (Perkin Elmer, Shelton, CT, USA) operating in electron impact mode at 70 eV and a scan range of 40–300 atomic mass unit (amu). The gas chromatograph was equipped with an Zebron ZB-Wax (30 m × 0.25 mm × 0.25 µm) column (Phenomenex. Torrance, CA, USA). The GC/MS injector and detector were set at 250 °C and 200 °C, respectively. The column temperature was initially held at 35 °C for 5 min, then increased at 5 °C/min to 200 °C, and held for 5 min. GC/MS analytical samples were run in triplicate and VC were identified based on comparisons of retention indices and mass spectra with literature data and by comparison with the NIST-14 Mass Spectral Library (US National Institute of Standards and Technology).

2.4.2. Extraction of Lipids from Chia Oil Based Mayonnaise

The determination of degradation products in the mayonnaise samples was carried out by extracting the lipids using chloroform/methanol (1:1 v/v) as solvent. Two extractions were made from each sample and the extracted lipids were stored at $-20\,^\circ$C under nitrogen until analysis.

2.4.3. Measurement of Degradation Products

Oil samples extracted from each treatment were used to measure, using official methods, acidity (AI) and peroxide indexes (PI), dienes and trienes conjugates (k_{232}–k_{270}), and concentrations of secondary oxidation products through the p-anisidine value (p-AnV). The TOTOX value (2PI + p-AnV)—which provides an idea of the total oxidative state of a system—was also calculated.

2.5. Statistical Analysis

All experiments were performed in triplicate. The results were depicted in the form of average and standard deviation. Comparisons between treatments were performed by means of ANOVA test using INFOSTAT/Professional 2014 software (FCA-UNC, Argentina). Normality of data was tested using Shapiro–Wilk test. Results giving p values < 0.05 were considered significantly different.

3. Results and Discussion

3.1. Physical Properties of Mayonnaise

The addition of PSE did not affect the pH and a_w of the chia oil-based mayonnaise; their values were 5.73 ± 0.05 and 0.901–0.925, respectively. Table 1 summarized the color and particle size results. The incorporation of PSE caused a change in the color of the mayonnaise compared to the control, but it did not affect the mean particle size. The parameter with the highest variation was a*, which indicated an increase in the red tones of the samples when PSE was incorporated into the formulation. Both parameters, *WI* and *YI*, are significantly different in both mayonnaise treatments.

Table 1. Color and particle size parameters of chia oil based mayonnaise without antioxidants (Control) and with peanut skin extract (PSE) (2 mg/g) at initial storage time.

Treatment	Color					Particle Size	
	L*	a*	b*	WI	YI	$d_{4,3}$ (µm)	Span
Control	67.01 ± 0.94 [b]	0.81 ± 0.19 [a]	31.10 ± 0.77 [b]	-26.3 ± 1.71 [a]	66.31 ± 1.99 [b]	1.29 ± 0.18	0.95 ± 0.03
PSE	39.19 ± 0.77 [a]	4.55 ± 0.15 [b]	12.38 ± 0.63 [a]	2.06 ± 1.40 [b]	45.12 ± 1.62 [a]	1.22 ± 0.02	1.13 ± 0.14

Different letters in the same column indicates significative differences ($p > 0.05$).

Regarding the texture of the mayonnaise, firmness is an important parameter as it can affect sensory characteristics and applicability. However, the addition of PSE did not have an effect on the firmness (1.76 ± 0.03 in both treatments) or adhesiveness (17.27 ± 1.49).

3.2. Degradation Products during Mayonnaise Storage

The PSE used in this study, composed mainly of procyanidin and proanthocyanidin oligomers [3], has been found to display strong antioxidant capacities as measured by different reducing power and radical scavenging methods [6,7]. However, solubility is also important for characterizing the antioxidant activity of phenolic natural extracts. Partition coefficients (*log* P) of the PSE showed negative values (pH 3.5= -1.19; pH 5.5 = -1.21) [6]. But the absolute values were not as high as might be predictable for an extract with high polyhydric phenols content. Therefore, PSE phenolics could cover a relatively wide range of *log* P, which in turn could favor their partial dissolution in oil. The information mentioned above could be significant in emulsion systems because the components of PSE have the potential to diffuse through the aqueous phase and reach the interface. This could alter the physical and chemical properties of the interface, potentially affecting the oil stability. Indeed, PSE demonstrated high antioxidant efficacy when tested in an acid model O/W emulsion made with chia oil, Tween 80 and acid buffer, maintained at 40 °C for 15 days [6].

Table 2 shows values of oxidative degradation products. In both treatments, at the initial storage time (t = 0), the peroxide index (PI) was 1.36 meq O_2/kg oil; at the final storage time in control it reached a value of 74.7 meq O_2/kg oil. This value was significantly higher than that reached in PSE-added mayonnaise (13.4 meq O_2/kg oil). On the other hand, the acidity (AI) did not show significant variations ($p > 0.05$) among the different times in each treatment, except for the last month where the differences between the PSE and control treatments were significant (Table 2).

Table 2. Peroxide Index (PI), k_{232}, k_{270}, Acidity Index (AI), p-Anisidine and TOTOX value of lipids extracted from chia oil based mayonnaise without antioxidants (Control) and with peanut skin extract (PSE) (2 mg/g) at different storage (25 °C) time.

Mayonnaise Treatment	Month	PI (meq O_2/kg Chia Oil)		K_{232}		K_{270}		AI (mg KOH/ kg Chia Oil)		P-Anisidine Value		TOTOX Value	
Control	0	1.36 ^{Aa}	± 0.43	2.28 ^{Aa}	± 0.05	0.60 ^{Aa}	± 0.05	1.98 ^{Aa}	± 0.02	ND		2.72 ^{Aa}	± 0.70
	2	24.38 ^{Bb}	± 2.37	3.90 ^{Bb}	± 0.43	0.55 ^{Ba}	± 0.01	1.98 ^{Aa}	± 0.16	0.19 ^{a} ± 0.08		48.97 ^{Bb}	± 4.82
	4	50.41 ^{Bc}	± 1.02	6.66 ^{Bc}	± 0.02	0.83 ^{Bb}	± 0.01	1.91 ^{Aa}	± 0.23	2.13 ^{b} ± 0.02		102.95 ^{Bc}	± 2.07
	6	74.69 ^{Bd}	± 7.31	10.05 ^{Bd}	± 0.70	1.34 ^{Bc}	± 0.04	2.37 ^{Bb}	± 0.11	3.68 ^{Bc} ± 0.47		153.08 ^{Bd}	± 14.32
PSE	0	1.36 ^{Aa}	± 0.43	2.28 ^{Aa}	± 0.05	0.60 ^{Aa}	± 0.05	1.98 ^{Aa}	± 0.02	ND		2.72 ^{Aa}	± 0.70
	2	3.83 ^{Ab}	± 0.58	2.47 ^{Ab}	± 0.05	0.49 ^{Aa}	± 0.01	1.90 ^{Aa}	± 0.08	ND		7.67 ^{Ab}	± 1.17
	4	9.22 ^{Ac}	± 0.38	2.97 ^{Ac}	± 0.12	0.54 ^{Aa}	± 0.06	1.91 ^{Aa}	± 0.12	ND		18.44 ^{Ac}	± 0.76
	6	13.36 ^{Ad}	± 0.50	3.55 ^{Ad}	± 0.01	0.64 ^{Aa}	± 0.01	2.01 ^{Aa}	± 0.02	0.98 ^{A} ± 0.19		27.70 ^{Ad}	± 1.19

ND: no detected. Different letters indicate statistically significant differences ($p > 0.05$). Capital letters indicate differences between treatments in the same month and lower-case letters differences between months within the same treatment.

Oils that contain high amounts of polyunsaturated fatty acids (PUFAs), such as chia oil, presents a faster increase in conjugated dienes compared to oils or fats with lower levels of PUFAs. As a result, the increase in K_{232} during storage was noticeable in both treatments (Table 2), but it was much more prompt in the control mayonnaise, reaching a final value of 10.05, while in the PSE treatment, the final value was 3.55. On the other hand, K_{270} remained constant (about 0.6) throughout the storage in mayonnaise with PSE, and in the control from the fourth month they increased slightly until reaching a final value of 1.34.

Decomposition of fatty acid hydroperoxides generates a large variety of secondary oxidation products, which can be measured by pAnV. This index is indicative of aldehydic compounds, mainly 2-alkenals and 2,4-alkadienals. These compounds were not detected in PSE-added mayonnaise until sixth month with a value of 0.98. In the control treatment, the value was 0.19 at the second month, and increased to 3.68 at the end of the storage period (Table 2). The TOTOX index at this time showed a value of 153.08 in control mayonnaise, while in the PSE treatment, it was 27.70, six times less.

Furthermore, lipid oxidation can result in off-flavors, and various volatile compounds (VC) are responsible for this. Changes in selected VC found in the chia oil-based mayonnaise at 0 and 6 months are shown in Table 3. Autoxidized linolenic acid produce decatrienal methyl octanoate from the 9-hydroperoxide; 2,4-heptadienal from the 12-hydroperoxide; 3-hexenal and 2-pentenal from the 13-hydroperoxide; propanol and ethane from the 16-hydroperoxide [10]. In this study, only 2,4-heptadienal and 3,5-octadiene-2-one were quantified due to the method detection limits. These compounds have been previously reported as the predominant VC in a chia oil accelerated oxidation test [10]. In the control mayonnaise, the final concentration of 2,4-heptadienal was 3.75 µg/g, while in the PSE treatment, the value was approximately five times lower (0.83 µg/g). In contrast, 3,5-octadiene-2-one was not detected in the mayonnaise with PSE, while in the control treatment the value was 2.15 µg/g.

Table 3. Concentration of majority volatile compound (VC) from chia oil based mayonnaise without antioxidants (Control) and with peanut skin extract (PSE) (2 mg/g) at initial (t0) and final time (6 month) of storage (25 °C) assay.

Volatile Compounds (VC)	Concentration (μg VC/g Mayonnaise)						
	t0	Control			PSE		
		6 month/25 °C					
2.4 Heptadienal	ND	3.75[b]	±	0.44	0.83[a]	±	0.21
3.5-Octadiene-2-one	ND	2.15	±	0.84	ND		

ND: no detected. Different letters in the same row indicates significative differences ($p > 0.05$).

4. Conclusions

Despite the unfavorable conditions for food preservation (water activity 0.9, pH 5.73, air-oxygen, and 25 °C), the chia oil-based mayonnaise with 2 mg/g of a natural antioxidant extract from peanut skin (PSE) was highly stable for six months. Therefore, the results of this storage test demonstrate that PSE phenolic compounds have a high potential to reduce primary and secondary oxidation products in this emulsion food system.

Although the effective concentration of PSE could be much higher than those usually used for synthetic antioxidants (0.1–0.2 mg/g), it should be noted that worldwide food regulations recognize natural antioxidants as *GRAS* additives; so, there are no regulations that limit their use.

Author Contributions: Conceptualization, R.M.B., D.M.M. and M.L.M.; methodology, R.M.B., A.C.R.-R. and M.L.M.; validation, R.M.B., A.C.R.-R. and M.L.M.; formal analysis, R.M.B., A.C.R.-R., A.R.V. and M.L.M.; investigation, R.M.B., A.C.R.-R. and M.L.M.; resources, R.M.B., D.M.M., P.D.R. and M.L.M.; data curation, R.M.B., A.C.R.-R. and M.L.M.; writing—original draft preparation, R.M.B.; writing—review and editing, D.M.M. and M.L.M.; visualization, R.M.B., A.C.R.-R., D.M.M., P.D.R., A.R.V. and M.L.M.; supervision, R.M.B. and M.L.M.; project administration, R.M.B., D.M.M., P.D.R. and M.L.M.; funding acquisition, R.M.B., D.M.M., P.D.R. and M.L.M. All authors have read and agreed to the published version of the manuscript.

Funding: This research was funded by grant Ia ValSe-Food (119RT0567), MinCyT-PIOBio (Decreto N° 349/2022-Res. Mi. N° 00000043/2022), CONICET (PIBAA-2021-28720210100284CO), ANPCyT (PICT-2020-SERIEA-01656) and SeCyT-UNC, Argentina.

Institutional Review Board Statement: Not applicable.

Informed Consent Statement: Not applicable.

Data Availability Statement: The data presented in this study are available on request from the corresponding author.

Acknowledgments: The authors would also like to acknowledge the Iberoamerican Project CYTED 119RT0567.

Conflicts of Interest: The authors declare no conflict of interest.

References

1. Jacobsen, C. Oxidative Stability and Shelf Life of Food Emulsions. In *Oxidative Stability and Shelf Life of Foods Containing Oils and Fats*; Hu, M., Jacobsen, C., Eds.; Elsevier Inc.: Amsterdam, The Netherlands, 2016; pp. 287–312. [CrossRef]
2. Wang, W.; Xiong, P.; Zhang, H.; Zhu, Q.; Liao, C.; Jiang, G. Analysis, occurrence, toxicity and environmental health risks of synthetic phenolic antioxidants: A review. *Environ. Res.* **2021**, *201*, 111531. [CrossRef] [PubMed]
3. Bodoira, R.M.; Cittadini, M.C.; Velez, A.; Rossi, Y.; Montenegro, M.; Martínez, M.L.; Maestri, D.M. An overview on extraction, composition, bioactivity and food applications of peanut phenolics. *Food Chem.* **2022**, *381*, 132250. [CrossRef] [PubMed]
4. Lorenzo, J.M.; Munekata, P.E.S.; Sant'Ana, A.S.; Carvalho, R.B.; Barba, F.J.; Toldra, F.; Trindade, M.A. Main characteristics of peanut skin and its role for the preservation of meat products. *Trends Food Sci. Technol.* **2018**, *77*, 1–10. [CrossRef]
5. Verstraeten, S.V.; Hammerstone, J.F.; Keen, C.L.; Fraga, C.G.; Oteiza, P.I. Antioxidant and membrane effects of procyanidin dimers and trimers isolated from peanut and cocoa. *J. Agric. Food Chem.* **2005**, *53*, 5041–5048. [CrossRef] [PubMed]

6. Bodoira, R.; Martínez, M.; Velez, A.; Cittadini, M.C.; Ribotta, P.; Maestri, D. Peanut skin phenolics obtained by Green solvent extraction: Characterization and antioxidant activity in pure chia oil and chia oil in water (*O/W*) emulsion. *J. Sci. Food Agric.* **2021**, *102*, 2396–2403. [CrossRef] [PubMed]

7. Bodoira, R.M.; Rossi, Y.; Montenegro, M.; Maestri, D.M.; Velez, A. Extraction of antioxidant polyphenolic compounds from peanut skin using water-ethanol at high pressure and temperature conditions. *J. Superc. Fluids.* **2017**, *128*, 57–65. [CrossRef]

8. Alizadeh, L.; Abdolmaleki, K.; Nayebzadeh, K.; Shahin, R. Effects of tocopherol, rosemary essential oil and *Ferulago angulata* extract on oxidative stability of mayonnaise during its shelf life: A comparative study. *Food Chem.* **2019**, *285*, 46–52. [CrossRef] [PubMed]

9. Miguel, G.A.; Jacobsen, C.; Prieto, C.; Kempen, P.J.; Lagaron, J.M.; Chronakis, J.S.; García-Moreno, P.J. Oxidative stability and physical properties of mayonnaise fortified with zein electrosprayed capsules loaded with fish oil. *J. Food Eng.* **2019**, *263*, 348–358. [CrossRef]

10. Hyojik, J.; Inhwan, K.; Sunghyeon, J.; Jihyun, L. Oxidative stability of chia seed oil and flax seed oil and impact of rosemary (*Rosmarinus officinalis* L.) and garlic (*Allium cepa* L.) extracts on the prevention of lipid oxidation. *Appl. Biol. Chem.* **2021**, *64*, 2–16. [CrossRef]

Proceeding Paper

Physicochemical and Nutritional Characterization of Paraguayan Organic *Moringa oleifera* Leaves as a Food Ingredient [†]

L. Mereles [1,*], M. L. Castelló [2], P. Piris [1], R. Villalba [1], E. Coronel [1], S. Caballero [1], J. López [1], A. Suárez [1], M. D. Ortolá [2], C. Figueras [3] and E. A. Gómez [2]

[1] Departamento Bioquímica de Alimentos, Facultad de Ciencias Químicas, Universidad Nacional de Asunción, Dirección de Investigaciones, San Lorenzo 1055, Paraguay; ppiris@qui.una.py (P.P.); rvillalba@qui.una.py (R.V.); ecoronel@qui.una.py (E.C.); scaballero@qui.una.py (S.C.); jlopez@qui.una.py (J.L.); adecia.suarez@gmail.com (A.S.)

[2] Food Engineering Research Institute (FoodUPV), Universitat Politècnica de València, Camino de Vera s/n, 46022 Valencia, Spain; mcasgo@upvnet.upv.es (M.L.C.); mdortola@tal.upv.es (M.D.O.); ergoval@etsiamn.upv.es (E.A.G.)

[3] Asociación Paraguay Orgánico, Asunción 1372, Paraguay; mcf@yguamoringa.com

* Correspondence: lauramereles@qui.una.py

[†] Presented at the V International Conference la ValSe-Food and VIII Symposium Chia-Link, Valencia, Spain, 4–6 October 2023.

Citation: Mereles, L.; Castelló, M.L.; Piris, P.; Villalba, R.; Coronel, E.; Caballero, S.; López, J.; Suárez, A.; Ortolá, M.D.; Figueras, C.; et al. Physicochemical and Nutritional Characterization of Paraguayan Organic *Moringa oleifera* Leaves as a Food Ingredient. *Biol. Life Sci. Forum* **2023**, 25, 18. https://doi.org/10.3390/blsf2023025018

Academic Editors: Claudia M. Haros and Loreto Muñoz

Published: 13 October 2023

Abstract: Knowledge of the physicochemical characteristics of organic moringa leaves, such as the color, the moisture control and water activity (aw) for biological control, as its basic composition, provides valuable starting data for its improvement and quality control through the selection of key parameters for the conservation of its multiple nutritional and antioxidant qualities in organic production systems adapted to local conditions. Objective: Describe the physicochemical characteristics and composition of macro- and micronutrients of *Moringa oleifera* leaves from a vivarium under controlled conditions and evaluate the antioxidant potential of dried moringa leaves from organic production in Paraguay under experimental conditions. A systematic sampling of moringa leaves from a vivarium was carried out. For color analysis, the CIEL*a*b* coordinates were determined. For physicochemical characteristics' analysis and macro- and micronutrient composition, official methods were used, while Total Phenolics Compounds (TPC) were determined by the Folin Ciocalteau method, and the Total Antioxidant Capacity (TAC) was determined by the antioxidant inhibition test (Radical ABTS+). Under experimental conditions, the dried moringa leaves under experimental conditions presented a light green color and low levels of humidity (8.1 ± 0.4%) and a_w (0.59), with high protein levels (27%), micronutrients such as minerals (Ca, Mg, P, Fe, Zn), vitamins B1, B2, Vit. C, TPC (5051 ± 168 g GAE/100 g) and antioxidants (468 ± 109 mMTEAC/g). Moringa leaves from Paraguay represent a source of micronutrients in the diet, and can be applied as ingredients in different culinary preparations within the framework of a healthy and sustainable diet, in accordance with the Sustainable Development Goals (SDGs).

Keywords: composition; characterization; *Moringa oleifera*; organic; nutrition; micronutrients

1. Introduction

The growing global population requires the development of efficient, inclusive, resilient and sustainable food systems. This compels innovation in the development of new approaches to promote and consume nutritious foods, adapted to the conditions of climate change, while respecting the environment and the traditions of vulnerable populations [1]. This challenge also involves the resilience of livelihoods and the revaluation of underutilized species with ancestral uses. In this context, *Moringa oleifera* leaves have been recognized since ancient times as a source of protein and other important nutrients such as vitamins and minerals [2]. Its cultivation is known for its ability to produce a high yield with a less carbon-intensive footprint compared to other crops [3]. It adapts well to warm climates with a limited water supply, making it a potential alternative crop in Paraguay [4]. Understanding physicochemical characteristics such as its leaf color, and implementing biological controls, such as humidity and water activity, along with its basic composition, provides valuable initial data for its enhancement and quality control [5]. This can be achieved through the selection of key parameters to conserve its numerous nutritional and antioxidant qualities within organic production systems tailored to local conditions. The aim of the work was to describe the physicochemical characteristics of moringa leaves from organic production in Paraguay from a nursery under controlled conditions, as well as their nutritional properties and antioxidant potential, based on their percentage composition, minerals, water-soluble vitamins, total polyphenols compounds and total antioxidant capacity (TAC).

2. Materials and Methods

2.1. Sampling

A systematic sampling of moringa leaves from a nursery with an organic fertilization system was carried out in Paraguay. For the analytical determination of nutrients and antioxidants, fresh moringa leaves were dried on the same day as harvest using a convective drying system in trays (50 °C-8 h) in an conventional oven [4].

2.2. Physical and Chemical Characterization

For the color analysis, the CIEL*a*b* coordinates were determined using a Spectrophotometer (TS7600 BOYN Co., Hangzhou, China) with a reference illuminant D65 and a 2° observer. Physicochemical characteristics and were composition of macro- and micronutrients as minerals and vitamins were determined using the official methods of the Association of Official Agricultural Chemists AOAC [6]. For vitamin C, the spectrofluorometric AOAC method 967.22 was used; minerals (P, Ca, Fe, Zn, Mg, Cu) were determined by the atomic absorption spectrophotometry AOAC method 975.03; TPC content was determined by the by Folin–Ciocalteau method [7] and the total antioxidant capacity was analyzed by ABTS$^{\bullet+}$ radical cation-based assay (2,2′-azino-bis(3-ethylbenzothiazoline-6-sulfonic acid) [8].

3. Results

The fresh moringa leaves had 14.5% moisture on the day of harvest; when they were dried, this decreased to 8.05%, which makes the resulting product non-perishable. The color results obtained in fresh and dry leaves are presented in Table 1. Under the air-drying conditions used, it was observed that storage leads to a significant alteration in the "b" value, while this is not significant for the "L" and "b" values (Student t test, $p < 0.05$). This results in a minor shift in the green hue, which is imperceptible to the naked eye.

Table 1. Physicochemical characterization of organic dry leaves of *Moringa oleifera* Lam.

Samples	Color			Moisture	aw
	L*	a*	b*	g/100 g	-
Dry leaf 1	56.66 ±0.15 [a]	−5.26 ± 0.020 [a]	20.97 ± 0.13 [a]	9.10 ± 0.10 [a]	0.58 ± 0.01 [a]
Fresh leaf 1	57.09 ±0.89 [a]	−6.10 ± 0.19 [b]	21.46 ± 0.29 [a]	14.50 ± 0.10 [b]	0.98 ± 0.01 [b]

L* indicates the brightness, and a* and b* are the chromatic coordinates. Results are expressed as mean ± standard deviation. aw: water activity. Values with the same letters as columns indicate that there is no significant difference (Student t test, $p < 0.05$).

Composition of Dry M. oleifera Leaves

The centesimal composition of the dried moringa leaves from organic production showed high protein content (26.77 ± 0.52 g/100 g), as well as high levels of dietary fiber (21.54 ± 0.30 g/100 g) and total carbohydrates (20.15 ± 0.47 g/100 g), where 12.49 ± 0.91 g/100 g corresponds to soluble sugars. A low lipid content (7.16 ± 0.13 g/100 g) was observed

(Table 2). The high ash content (10.64 ± 0.07 g/100 g) demonstrates an equally important contribution of minerals, including calcium (2620 ± 210 mg/100 g) and magnesium (362 ± 55 mg/100 g). It also provides significant amounts of phosphorus, iron, zinc, copper and manganese, with a low sodium content (Table 2). Regarding the content of antioxidants and vitamin micronutrients, high concentrations of total polyphenols (5051 ± 168 g GAE/100 g) and total antioxidant capacity (468 ± 109 mM TEAC/g) were observed in the dry leaves, as well as a good supply of vitamin C (23.54 ± 3.51 g/100 g), and riboflavin (23.54 ± 3.51 mg/100 g).

Table 2. Centesimal composition, caloric value, mineral content, water-soluble vitamins and antioxidants on organic dry leaves of *Moringa oleifera* Lam. harvested in Paraguay.

Centesimal Composition	Mean ± SD	Minerals	Mean ± SD
Moisture (g/100 g)	8.05 ± 0.39	Calcium (mg/100 g)	2620 ± 210
Proteins (g/100 g)	26.77 ± 0.52	Phosforus (mg/100 g)	367 ± 1
Total carbohydrates (g/100 g)	20.15 ± 0.47	Magnesium (mg/100 g)	362 ± 55
Sugars (g/100 g)	12.49 ± 0.91	Iron (mg/100 g)	21.87 ± 6.73
Total lipids (g/100 g)	7.16 ± 0.13	Sodium (mg/100 g)	12.26 ± 2.42
Ash (g/100 g)	10.64 ± 0.07	Zinc (mg/100 g)	7.07 ± 0.38
Dietary fiber (g/100 g)	21.54 ± 0.30	Potasium (mg/100 g)	4.23 ± 1.35
Valor calórico (Kcal/100 g)	252 ± 3	Cupper (mg/100 g)	3.42 ± 0.41
		Manganese (mg/100 g)	3.00 ± 0.19
Vitamins		**Antioxidants**	
Ascorbic acid (mg/100 g)	47.00 ± 3.17	Total phenolics compounds (mg GAE/100 g)	5051 ± 168
Riboflavin (mg/100 g)	35.81 ± 5.46		
Tiamin (mg/100 g)	0.77 ± 0.02	ABTS (mM TEAC/g)	468 ± 109

Results are expressed as mean ± standard deviation, g; grams, mg; miligrams, Kcal; kilocalories, GAE; galic acid equivalents, mMTEAC; milimoles equivalents of TROLOX.

4. Discussion

One of the most important sensory parameters for the acceptance of dry foods such as moringa leaf powder is color, which is essential for the acceptance of the product. Under the drying test treatment used, a darkening of the leaves was observed, which, due to the high content of beta carotene (yellowish color), are sensitive to heat, light and oxygen. Oven-drying treatments at 50 can also lead to significant color variations due to the carbohydrate contents. Low-molecular-weight carbohydrates are generally lost during heating, which can occur due to the rapid caramelization of the sugar at high temperatures. These variations are usually not significant when processes such as microwave-drying or freeze-drying are used, which increase production costs [5]. The effect of the drying treatments on the protein content is not significant from 40 to 60 degrees and the protein, carbohydrate and fat contents of dried Moringa leaves (4–11% moisture) increase by 3–4 times compared to fresh leaves due to the increase in dry matter [9]. The results observed in organic moringa leaves harvested in Paraguay show similar values to those reported in the literature, with a high contribution of proteins, carbohydrates and minerals. Likewise, the contribution of antioxidants and micronutrients stands out in the evaluated samples [10]. The potential nutritional value of dried moringa leaves from organic production in Paraguay suggests that they are an interesting alternative, which could be inserted as an ingredient in value-added products in formulations that are acceptable to consumers as a dietary supplement. Sensory evaluation studies on foods with dried moringa leaves and their nutritional contribution are necessary to insert them into the regional diet.

5. Conclusions

The dried, organically produced moringa leaves that were analyzed have high levels of vegetable protein, micronutrients such as minerals, vitamins, and other compounds of interest such as antioxidants. Moringa leaves in Paraguay can represent a source of micronutrients in the diet and be applied as ingredients in different culinary preparations within the framework of a healthy and sustainable diet, in accordance with the Sustainable Development Goals.

Author Contributions: Conceptualization, L.M. and M.L.C.; methodology, M.L.C. and L.M.; software, E.C.; validation, M.D.O., M.L.C. and R.V.; formal analysis, A.S., P.P. and E.A.G.; investigation, R.V.; investigation analysis, J.L.; resources, M.L.C., L.M. and S.C.; data curation, L.M.; writing—original draft preparation, L.M.; writing—review and editing, C.F.; visualization, M.D.O.; supervision, L.M. and M.L.C.; project administration, L.M. and M.L.C. All authors have read and agreed to the published version of the manuscript.

Funding: This work was supported by grant Ia ValSe-Food-CYTED (119RT0567) and was funded by the project MORNUPAY (Ref. AD2115- Centre for Development Cooperation (CCD)-Universitat Politècnica de València, Spain) and A. Gómez has been awarded with a grant from the MERIDIES-Cooperation 2023 Programme of the Centre for Development Cooperation (CCD) of the UPV, Spain.

Institutional Review Board Statement: Not applicable.

Informed Consent Statement: Not applicable.

Data Availability Statement: Not applicable.

Conflicts of Interest: The authors declare no conflict of interest.

References

1. FAO. *Climate Change, Biodiversity and Nutrition Nexus. Evidence and Emerging Policy and Programming Opportunities*; FAO: Rome, Italy, 2021; ISBN 978-92-5-134920-5.
2. Ribaudo, G.; Povolo, C.; Zagotto, G. Moringa oleifera Lam.: A Rich Source of Phytoactives for the Health of Human Being. In *Studies in Natural Products Chemistry*; Elsevier: Amsterdam, The Netherlands, 2019; Volume 62, pp. 179–210, ISBN 978-0-444-64185-4.
3. Kim, Y.-J.; Kim, H.S. Screening Moringa Species Focused on Development of Locally Available Sustainable Nutritional Supplements. *Nutr. Res. Pract.* **2019**, *13*, 529. [CrossRef] [PubMed]
4. Povolo, C.; Mereles, L.; Fretes, N.; Caballero, S.; Alvarenga, N.; Zagotto, G.; Rumignani, A.; Neresini, M. Aqueous extracts of Paraguayan Moringa (Moringa oleifera Lam.), as a source of direct and indirect antioxidants for nutraceutical applications. *Int. J. Food Sci. Nutr.* **2019**, *4*, 170–176.
5. Ali, M.A.; Yusof, Y.A.; Chin, N.L.; Ibrahim, M.N. Processing of Moringa Leaves as Natural Source of Nutrients by Optimization of Drying and Grinding Mechanism. *J. Food Process Eng.* **2017**, *40*, e12583. [CrossRef]
6. Horwitz, W.; Chichilo, P.; Reynols, H. *Official Methods of Analysis of the Association of Official Analytical Chemists*, 17th ed.; AOAC: Gaithersburg, MA, USA, 2000.
7. Singleton, V.L.; Rossi, J.A. Colorimetry of Total Phenolics with Phosphomolybdic-Phosphotungstic Acid Reagents. *Am. J. Enol. Vitic.* **1965**, *16*, 144–158. [CrossRef]
8. Re, R.; Pellegrini, N.; Proteggente, A.; Pannala, A.; Yang, M.; Rice-Evans, C. Antioxidant Activity Applying an Improved ABTS Radical Cation Decolorization Assay. *Free Radic. Biol. Med.* **1999**, *26*, 1231–1237. [CrossRef]
9. Deepak, S.; Suryawanshi, O.P.; Patel, S.; Jaiswal, A. Nutritive Evaluation of Fresh Moringa oleifera Leaves in Chhattisgarh Region. *Int. J. Chem. Stud.* **2021**, *9*, 1184–1188. [CrossRef]
10. Okiki, P.A.; Osibote, I.A.; Balogun, O.; Oyinloye, B.E.; Idris, O.; Olufunke, A.; Asoso, S.O.; Olagbemide, P.T. Evaluation of Proximate, Minerals, Vitamins and Phytochemical Composition of Moringa oleifera Lam. Cultivated in Ado Ekiti, Nigeria. *Adv. Biol. Res.* **2015**, *9*, 436–443.

Proceeding Paper

Development of a Nutritional Drink Based on Kurugua Wholemeal Flour as a Source of Minerals and Amino Acids [†]

Eva Coronel [1], Marcela L. Martínez [2,3], Edgardo Calandri [2,4], Rocío Villalba [1], Alexis Ortiz [1], Silvia Caballero [1] and Laura Mereles [1,*]

[1] Departamento Bioquímica de Alimentos, Dirección de Investigaciones, Facultad de Ciencias Químicas, Universidad Nacional de Asunción, San Lorenzo P.O. Box 1055, Paraguay; ecoronel@qui.una.py (E.C.); rvillalba@qui.una.py (R.V.); ortizalexiis4@gmail.com (A.O.); scaballero@qui.una.py (S.C.)

[2] Instituto de Ciencia y Tecnología de los Alimentos, Facultad de Ciencias Exactas, Físicas y Naturales (ICTA-FCEFyN)—Universidad Nacional de Córdoba (UNC), Córdoba 5000, Argentina; marcela.martinez@unc.edu.ar (M.L.M.); edgardo.calandri@unc.edu.ar (E.C.)

[3] Instituto Multidisciplinario de Biología Vegetal (IMBIV), Consejo Nacional de Investigaciones Científicas y Técnicas (CONICET), Universidad Nacional de Córdoba (UNC), Córdoba 5000, Argentina

[4] Instituto de Ciencia y Tecnología de los Alimentos Córdoba (ICYTAC), Universidad Nacional de Córdoba (UNC), Córdoba 5000, Argentina

[*] Correspondence: lauramereles@qui.una.py

[†] Presented at the V International Conference la ValSe-Food and VIII Symposium Chia-Link, Valencia, Spain, 4–6 October 2023.

Abstract: Adequate intake of mineral nutrients and amino acids is essential for nutrition in the Western diet, where deficiencies of minerals such as Zn, Fe, and good quality proteins are highly prevalent in vulnerable populations in Latin America and the Caribbean (LAC). The aim of the present study was to evaluate the mineral and amino acid content of Kurugua (KWF) wholemeal flour and a derived product (9% K w/v, 0.8% chia oil (w/v), and 1% sweetener v/v). Proteins were analyzed by Microkjeldhal, minerals Na, K, P, Ca, Mg, Zn, Cu, Mn, and Fe by AOAC, and amino acid profile by HPLC-DAD methods. KWF presented a high content of Mg and Zn (207.63 ± 5.27 and 15.76 ± 0.04 mg/100 g, respectively). A KWF-based drink provides 5.05 mg Zn/100 g, equivalent to 46% of the recommended daily intake (RDI) in a 200 mL serving of the beverage. Amino acids (glutamic acid + glutamine) and (aspartic acid + asparagine) were the most abundant in KWF (112.2 and 245 mg of aa/g protein, respectively), with 18.0% of total protein. A serving of KWF-based drink contains about 3.02 g of protein and the essential amino acids Hys, Val, Met, Phe, Ile, Leu, and Lys (31.6, 213, 198, 89.3, 186, 3.7, and 194.3 mg AA/200 mL, respectively). The ready-to-drink Kurugua drink takes full advantage of the wholemeal flour of an indigenous fruit such as Kurugua, providing a source of zinc and an adequate amount of essential amino acids and expanding the supply of healthy products within the framework of food safety.

Keywords: *Sicana odorifera*; kurugua; amino acids; minerals; zinc

Citation: Coronel, E.; Martínez, M.L.; Calandri, E.; Villalba, R.; Ortiz, A.; Caballero, S.; Mereles, L. Development of a Nutritional Drink Based on Kurugua Wholemeal Flour as a Source of Minerals and Amino Acids. *Biol. Life Sci. Forum* **2023**, 25, 19. https://doi.org/10.3390/blsf2023025019

Academic Editors: Claudia M. Haros, Loreto Muñoz and María Dolores Ortolá

Published: 19 October 2023

1. Introduction

The accelerated destruction of natural habitats due to land use changes, particularly for agriculture and urbanization, leads to the loss of genetic diversity in ecosystems and the displacement of traditional and ancestral crops such as chia and Kurugua [1]. These crops, though undervalued, have the potential to enhance dietary diversity and people's quality of life. However, their nutritional properties and value chains are not well understood. This scenario underscores the need to recognize the nutritional potential of these crops to improve food security [2].

Adequate intake of mineral nutrients and amino acids is essential for nutrition in the Western diet, where deficiencies of minerals such as Zn, Fe, and good quality proteins are highly prevalent in vulnerable populations in Latin America and the Caribbean (LAC) [3].

The objective of the present study was to evaluate the mineral and amino acid content of Kurugua (KWF) wholemeal flour and a derived product. This assessment aims to determine its nutritional value and potentially propose an alternative nutritious beverage.

2. Materials and Methods

2.1. Sample

The whole flour of Kurugua (KWF) was obtained from nine whole fruits of *Sicana odorifera*, as described previously [4]. From this flour, a derived product was created, which was a plant-based beverage with a formulation of 9% KWF *w/v*, 0.8% chia oil (*w/v*), and 1% sweetener *v/v* in water. The beverage mixture was prepared using a colloid mill.

2.2. Minerals

For mineral analysis such as Na, K, Ca, Mg, Fe, Cu, Zn, and Mn, the atomic absorption AOAC method 975.03 was used by triplicate [5]. An AA 6300 Shimadzu (Kyoto, Japan) spectrophotometer was employed. The phosphorus content was determined using the colorimetric AOAC method 935.45. For each mineral, a calibration curve was made using standard solutions (Merck, Darmstadt, Alemania).

2.3. Proteins and Amino Acids Profile

The protein analysis was determined by AOAC Method 920.152 [5], employing the conversion factor of 6.25 from N2 to proteins. For amino acid (AA) analysis, defatted samples (500 mg) were hydrolysed following the standard AOCS procedure [6]. Hydrolysed samples were filtered using a 0.45 μm membrane filter and then submitted to pre-column derivatization as described previously [7]. Amino acid derivatives were analyzed by reversed-phase HPLC (PerkinElmer, Shelton, CT, USA) under conditions used previously [7]. Identification was carried out by means of authentic AA standards (Sigma-Aldrich, St. Louis, MO, USA). Quantification was performed by the external standard method. The linearity of the response was verified by fitting the line results of each one of the AA individuals (five standard solutions with known concentrations), covering the concentration range from 10 to 100 ppm. Calibration curves from the tested AA standards showed linearity regression coefficients (R2) ranging between 0.96 and 0.99.

2.4. Statistical Analyses

The data were recorded in an Excel spreadsheet and analyzed in the statistical program GraphPad prism 5.0 (GraphPad Software Inc., CA, USA). Student's t ($p \leq 0.05$) was used to determine the significant differences.

3. Results

Table 1 displays the mineral results for both KWF and the derived product, as well as in a 200 mL portion of the beverage. Notably, Mg and Zn stand out in both samples, with Mg and Zn values of 207.63 and 15.76 mg/100 g in KWF and 222.94 and 30.20 mg/100 g in the product on a dry basis; contributing 37.27 and 5.05 mg of the respective minerals in a 200 mL serving of the beverage. The protein content in KWF was 18.00 ± 0.29 g/100 g, in the beverage on a dry basis, there was 16.23 ± 0.18 g/100 g, and in a 200 mL portion of the beverage, there was 2.71 g of protein.

Table 2 contains the results of the amino acid profile. The derived product contained 16.23 g/100 g protein on a dry basis and 1.51 g/200 mL serving in the reconstituted drink, which would provide eight essential amino acids with the exception of tryptophan.

Table 1. Minerals in Whole Kurugua Flour (KWF) and Its Derived Product.

Mineral	Kurugua Wholemeal Flour (mg/100 g)	Derived Product Dry Basis (mg/100 g)	Derived Product in 200 mL (mg/200 mL)	% of RDI per 200 mL
Na	13.09 ± 0.30 [a]	41.49 ± 13.96 [b]	8.59	0.4
K	1734 ±124 [a]	2638 ± 78.6 [b]	441.22	
Fe	6.85 ± 0.32 [a]	6.65 ± 0.41 [a]	1.11	6.2
Ca	58.99 ± 0.91 [a]	122.77 ± 8.06 [b]	20.52	1.6
Mg	207.63 ± 5.27 [a]	222.94 ± 17.30 [a]	37.27	8.9
Cu	2.68 ± 0.15 [a]	1.73 ± 0.05 [b]	3.56	395
Zn	15.76 ± 0.04 [a]	30.20 ± 2.28 [b]	5.05	46
Mn	1.11 ± 0.06 [a]	0.46 ± 0.06 [b]	0.08	3.4
P	559.24 ± 11.46 [a]	532.41 ± 12.82 [a]	43.71	3.5

Values are expressed as mean ± standard deviation (n = 3). Different lowercase letters in each row indicate a significant difference between the means, Student's *t*-test ($p \leq 0.05$).

Table 2. Amino Acid Profile in Whole Kurugua Flour (KWF) and Its Derived Product.

Amino Acids	Kurugua Wholemeal Flour (mg AA/g Protein)	Derived Product on a Dry Basis (mg AA/g Protein)	Derived Product in 200 mL (mg AA/Portion Serving)	Reference Composition (mg AA/g Protein) [a]
Ac. Asp+ Asn	112	112	304	
Ac. Glu + Gln	248	248	671	
Ser	21.1	21.9	57.6	
Hys	10.6	10.5	29.2	15
Gly	13.3	13.5	36.7	
Thr	35.0	35.1	95.2	
Pro	13.3	13.6	36.7	
Arg	20.6	20.3	55.9	
Ala	108	108	292	
Val	71.1	70.9	192	39
Met	65.6	65.6	178	22 *
Phe + Tyr	37.7	37.6	103	38 **
Ile	61.7	61.6	167	30
Leu	1.1	1.2	3.3	59
Lys	64.4	64.4	174	45

[a] Reference: Adapted with permission from Joint WHO/FAO/UNU Expert Consultation FAO/WHO/UNU [2007] [8]. * Corresponds to the protein composition reference value of Met + Cys (not determined) and ** to that of Phe + tyrosine.

4. Discussion

Significant differences in mineral content were found between Kurugua whole fruit flour and the plant-based beverage, except for Fe, Mg, and P. This is probably due to the contribution of these minerals from the Kurugua flour, which does not vary its content in the derived drink. Particularly noteworthy is the high content of Zn in both samples, contributing up to 46% of the Recommended Daily Intake (RDI) for Zn. Previous reports indicated Zn values of 42 mg/100 g and 2 mg/100 g in Kurugua pulp and seeds, respectively [9]. The values obtained in this study fall within this range.

The plant-based proteins in the samples were analyzed for their amino acid profiles, to evaluate their potential contribution as quality proteins. Based on the results and in accordance with the nutritional claims, a 200 mL serving of the vegetable drink would provide seven essential amino acids (histidine, threonine, arginine, valine, methionine, phenylalanine, isoleucine, leucine, and lysine), exceeding the amino acid levels of the reference food for good quality proteins [8]. However, for a food to be considered a source of protein, it must provide at least 6 g of protein per serving. This indicates the need to continue adjusting the formulation based on the functional properties of the product to obtain a protein drink; however, the profile of essential amino acids is interesting for this nutritional purpose. Pérez Venegas et al. [10] examined the amino acid profiles of commonly used plant-based beverage raw materials and reported lower amino acid values compared to those found in this study.

5. Conclusions

In this work, the mineral, protein, and amino acid content of Kurugua whole fruit flour and a derived plant-based beverage were analyzed. The beverage stands out for its zinc content and amino acid profile, positioning it as a valuable option for enriching the diet with essential nutrients. The utilization of the entire fruit, including the typically inedible skin, contributes nutrients to these products.

Author Contributions: Conceptualization, M.L.M., E.C. (Edgardo Calandri) and L.M.; methodology, M.L.M., E.C. (Edgardo Calandri) and L.M. software, E.C. (Eva Coronel); validation, E.C. (Eva Coronel) and L.M.; formal analysis, E.C. (Eva Coronel), A.O., R.V., A.O. and S.C.; investigation, E.C. (Eva Coronel), E.C. (Edgardo Calandri) and M.L.M.; resources, L.M., S.C.; data curation, L.M.; writing—original draft preparation, L.M.; writing—review and editing, M.L.M.; visualization, E.C. (Edgardo Calandri). and R.V.; supervision, L.M.; project administration, L.M. All authors have read and agreed to the published version of the manuscript.

Funding: This work was supported by grant Ia ValSe-Food-CYTED (119RT0567), COOPI (EuropeAid/154653/DD/ACT/Multi, UE), and Kurugua Poty Foundation, Paraguay.

Institutional Review Board Statement: Not applicable.

Informed Consent Statement: Not applicable.

Data Availability Statement: Not applicable.

Acknowledgments: This work was supported by grant Ia ValSe-Food-CYTED (119RT0567) and "Kurugua Poty" Foundation, Paraguay.

Conflicts of Interest: The authors declare no conflict of interest.

References

1. De Egea Elsam, J.; Céspedes, G.; Peña-Chocarro, M.d.C.; Mereles, F.; Rolón Mendoza, C. Recursos fitogenéticos del Paraguay: Sinopsis, Atlas y Estado de Conservación de los Parientes Silvestres de Especies de Importancia para la Alimentación y la Agricultura. *Rojasiana* **2018**, 232.
2. FAO. *Climate Change, Biodiversity and Nutrition Nexus. Evidence and Emerging Policy and Programming Opportunities*; FAO: Roma, Italy, 2021; ISBN 978-92-5-134920-5.
3. FAO; FIDA; OPS; PMA y UNICEF. *Panorama Regional de la Seguridad Alimentaria y Nutricional-América Latina y el Caribe 2022: Hacia una Mejor Asequibilidad de las Dietas Saludables*; FAO: Santiago, Chile, 2023. [CrossRef]
4. Correa, L.; Gamarra, M.; Sanabria, K.; Coronel, E.; Martínez, M.L.; Calandri, E.; Caballero, S.; Mereles, L. Obtaining Integral Kurugua Flour with Antioxidant Potential as Ingredient Foodstuffs. *Biol. Life Sci. Forum* **2022**, *17*, 22. [CrossRef]
5. Horwitz, W.; Chichilo, P.; Reynols, H. *Official Methods of Analysis of the Association of Official Analytical Chemists*, 17th ed.; AOAC: Gaithersburg, MA, USA, 2000.
6. Firestone, D. (Ed.) *Recommended Practices of the American Oil Chemist's Society AOCS*, 6th ed.; AOCS: Urbana, IL, USA, 2009.
7. Cittadini, M.C.; García-Estévez, I.; Escribano-Bailón, M.T.; Bodoira, R.M.; Barrionuevo, D.; Maestri, D. Nutritional and Nutraceutical Compounds of Fruits from Native Trees (Ziziphus Mistol and Geoffroea Decorticans) of the Dry Chaco Forest. *J. Food Compos. Anal.* **2021**, *97*, 103775. [CrossRef]
8. WHO/FAO/UNU. Protein and Amino Acid Requirements in Human Nutrition. Joint WHO/FAO/UNU Expert Consultation. *World Health Organ.* **2007**, *935*, 1–265. Available online: https://pubmed.ncbi.nlm.nih.gov/18330140/ (accessed on 27 September 2023).
9. Mereles, L.; Caballero, S.; Burgos-Edwards, A.; Benítez, M.; Ferreira, D.; Coronel, E.; Ferreiro, O. Extraction of Total Anthocyanins from Sicana Odorifera Black Peel Fruits Growing in Paraguay for Food Applications. *Appl. Sci.* **2021**, *11*, 6026. [CrossRef]
10. Pérez Venegas, L.S.; Restrepo Molina, D.A.; López Vargas, J.H. Características de las bebidas con proteína de soya. *Rev. Fac. Nac. Agron.* **2009**, *62*, 5165–5175.

biology and life sciences forum

Proceeding Paper

Formulation of Sustainable Biopolymer-Based Nanoparticles Obtained via Media Milling for Chia Oil Vehiculization in Pickering Emulsions [†]

María G. Bordón [1,2,3,*], Lucía López-Vidal [4], Nahuel Camacho [4], Marcela L. Martínez [2,3,5], María C. Penci [1,2,3], Cecilio Carrera-Sánchez [6], Víctor Pizones Ruiz Henestrosa [6], Santiago D. Palma [4] and Pablo D. Ribotta [1,2,3]

[1] Food Science and Technology Institute, Instituto de Ciencia y Tecnología de Alimentos Córdoba, Consejo Nacional de Investigaciones Científicas y Técnicas (ICYTAC, CONICET), Córdoba 5000, Argentina; cecilia.penci@unc.edu.ar (M.C.P.); pribotta@agro.unc.edu.ar (P.D.R.)

[2] Food Science and Technology Institute, Instituto de Ciencia y Tecnología de los Alimentos, Universidad Nacional de Córdoba (ICTA, UNC), Córdoba 5000, Argentina; marcela.martinez@unc.edu.ar

[3] Department of Industrial and Applied Chemistry, Facultad de Ciencias Exactas, Físicas y Naturales, Universidad Nacional de Córdoba (FCEFyN, UNC), Córdoba 5000, Argentina

[4] Research and Development Unit in Pharmaceutical Technology, Unidad de Investigación y Desarrollo en Tecnología Farmacéutica, Consejo Nacional de Investigaciones Científicas y Técnicas (UNITEFA, CONICET), Córdoba 5000, Argentina; l.lopezvidal@unc.edu.ar (L.L.-V.); nahuelc03@gmail.com (N.C.); sdpalma@unc.edu.ar (S.D.P.)

[5] Multidisciplinary Institute of Plant Biology, Instituto Multidisciplinario de Biología Vegetal, Consejo Nacional de Investigaciones Científicas y Técnicas (IMBIV, CONICET), Córdoba 5016, Argentina

[6] Chemical Engineering Department, Faculty of Chemistry, University of Seville, 41004 Seville, Spain; cecilio@us.es (C.C.-S.); vicpizrui@us.es (V.P.R.H.)

[*] Correspondence: maria.gabriela.bordon@mi.unc.edu.ar

[†] Presented at the V International Conference la ValSe-Food and VIII Symposium Chia-Link, Valencia, Spain, 4–6 October 2023.

Citation: Bordón, M.G.; López-Vidal, L.; Camacho, N.; Martínez, M.L.; Penci, M.C.; Carrera-Sánchez, C.; Pizones Ruiz Henestrosa, V.; Palma, S.D.; Ribotta, P.D. Formulation of Sustainable Biopolymer-Based Nanoparticles Obtained via Media Milling for Chia Oil Vehiculization in Pickering Emulsions. *Biol. Life Sci. Forum* **2023**, 25, 20. https://doi.org/10.3390/blsf2023025020

Academic Editors: Claudia M. Haros, Loreto Muñoz and María Dolores Ortolá

Published: 5 December 2023

Abstract: Sustainable corn starch nanoparticles were prepared using media milling to stabilize omega-3-rich Pickering emulsions based on chia oil. The milling conditions were as follows: 24 h (milling time), 0.4–0.6 mm (bead diameter), 1600 rpm (impeller speed), 30% (volume occupied by the grinding media), 7% w/v (starch concentration), and 0, 0.07 and 1% w/v of sodium dodecyl sulfate (SDS). Nanosuspensions containing 7% w/v of starch and the three concentrations of SDS were filtered, centrifuged, homogenized, and spray-dried to obtain redispersible powders. The particle size ranges were 2288 ± 211, 385 ± 21, and 278 ± 11 nm with 0, 0.07 and 1% w/v of SDS, respectively. The most stable backscattering profiles obtained during a period of one week were observed with 0.07 and 1% w/v of SDS. Therefore, the surface dilatational rheology of these particles adsorbed at chia oil/water interfaces was studied. A rapid decrease in the interfacial tension within 1 h was obtained with 1% w/v of SDS (down to 3 mN/m). Moreover, the most stable particle size after redispersion was obtained with the highest concentration of SDS. Finally, Pickering emulsions were prepared, and significant coalescence was observed with 0 and 0.07% w/v of SDS (within a few minutes). Nonetheless, in the presence of 1% w/v of SDS, oil droplets showed mean diameters and polydispersity indexes of 280.13 ± 4.60 nm and 0.35 ± 0.02, respectively, with no significant variations during storage for around 1 month. The results show that wet-stirred media milling can be applied to produce sustainable, new food-grade starch nanoparticles able to deliver bioactive compounds from chia oil.

Keywords: chia oil; corn starch; nanoparticle; Pickering emulsion; wet-stirred media milling

1. Introduction

The increased worldwide consumption of functional foods has boosted the design of new products through the incorporation of bioactive compounds. Chia oil (CO) is known for its rich content of α-linolenic acid (60–66%), with extensively reviewed health benefits [1]. However, the polyunsaturated fatty acids present make it prone to oxidation when exposed to environmental factors (air, light, and temperature). Hence, specific technological alternatives, such as Pickering emulsions, have been applied to stabilize omega-3-rich oils [2].

Pickering emulsions are those stabilized by micro and nanoparticles present at the interfacial area of two immiscible liquids. These dispersed systems possess several outstanding properties: greater physical stability, less toxicity, enhanced responses to external stimuli, and resistance to flocculation and coalescence [3]. Diverse biocompatible materials have been applied for the development of nanoparticles. Among them, starches of different botanical origins are recognized within the food industry as efficient stabilizers for Pickering emulsions [2,3].

The development of nanoparticles using physical top-down methodologies, such as wet-stirred media milling, offers several advantages: continuous operation, scalability, simple construction, enhanced size reduction velocity, and absence of organic solvents [4].

Based on the above context, the aim of this research was to obtain stable corn starch nanoparticles via wet-stirred media milling, in order to stabilize Pickering emulsions as delivery systems for chia oil. Compared with a previous work by our group [5], a higher oil load could be stabilized. Redispersible powders were prepared by spraying the nanosuspensions due to the enhanced conservation, storage, and transport of solid materials compared with the parent fluid suspensions.

2. Materials and Methods

2.1. Materials

Chia seed oil (CSO) (Nutracéutica Sturla SRL, Salta, Argentina) was obtained via cold pressing as described by Martínez et al. [6] with a pilot plant screw press (Komet Model CA 59 G, IBG Monforts, Mönchengladbach, Germany). On the other hand, corn starch was purchased from a local distributor (Distribuidora NICCO, Córdoba, Argentina); the stabilizer added during milling experiments was sodium dodecyl sulfate (SDS) (Sigma–Aldrich, San Luis, MO, USA). Other reagents were HPLC or analytical grade.

2.2. Milling Experimental Design

Corn starch nanoparticles were obtained via media milling using a NanoDisp® laboratory-scale mill (NanoDisp®, Córdoba, Argentina), which has been described in previous works [5,7].

Suspensions containing starch (7% w/v) and SDS in three different concentrations (0, 0.1, and 1% w/v) were milled according to Bordón et al. [5]. The milling conditions were the following: diameter of zirconia beads (0.4–0.6 mm); filling ratio with beads (30% v/v); impeller speed (1600 rpm); and milling time (24 h). The obtained nanosuspensions were subsequently filtered (200 ASTM screen, Zonytest, Buenos Aires, Argentina) (74 μm), centrifuged (13,000× g 40 min), and homogenized (18,000 rpm, 2 min, Ultraturrax homogenizer IKA T18, Janke & Kunkel GmbH, Staufen, Germany; subsequently, 1 cycle at 700 bar, in a high-pressure valve homogenizer, EmulsiFlex C5, Avestin, Ottawa, ON, Canada).

2.3. Characterization of Corn Starch Nanosuspensions

The obtained corn starch nanosuspensions were characterized according to average particle size (Z_{av}), polydispersity index (PDI) (immediately after milling and after storage for one month), and ζ-potential via dynamic light scattering at 25 °C (Nano Zetasizer, Malvern Instruments, Worcestershire, UK) [5]. Moreover, the rheological behavior was assessed with a controlled-stress rheometer, Physica MCR 301 Anton Paar (Physica Messtechnik, Ostfildern, Germany) [7].

2.4. Preparation and Characterization of Pickering Emulsions Stabilized with Corn Starch Nanoparticles

The nanosuspensions obtained in Section 2.2 were filtered, centrifuged, and blended with CSO (solids/CO ratio: 1/1 *w/w*) via high-speed homogenization (14,000 rpm, 3 min). Afterward, these coarse emulsions were passed through a high-pressure valve homogenizer (2 cycles, 700 bar, EmulsiFlex C5, Avestin, Ottawa, ON, Canada) to produce fine emulsions. The oil droplet size distribution, polydispersity index, and ζ-potential were measured as described in Section 2.3.

2.5. Spray Drying and Characterization of Powders

The nanosuspensions obtained in Section 2.2 were spray-dried to obtain redispersible powders. The spray drying process was performed in laboratory scale equipment, Büchi B-290 (Büchi Labortechnik AG, Flawil, Switzerland), according to Fu et al. [8]. The obtained powders were redispersed in Milli-Q water to determine the Z_{av} and the PDI as described in Section 2.3. The morphology of powders was assessed via scanning electron microscopy (SEM, LSM5 Pascal; Zeiss, Oberkochen, Germany). The transmittance profiles after redispersion were determined according to [9] (Vertical Scan Analyzer Quick Scan, Coulter Corp., Miami, FL, USA) and the interfacial tension at chia oil/water interfaces was measured with the pendant droplet method (TRACKER pendant droplet tensiometer IT Concept, Saint-Ouen-l'Aumône, France) as described in [10].

3. Results and Discussion

3.1. Characterization of Nanosuspensions and Pickering Emulsions

Nanomilling could modify the granular morphology and molecular structure in starch. Therefore, improved physicochemical properties are observed, such as low gelatinization temperature, enhanced paste viscosity and redispersion, and large specific surface area [2,3]. In order to obtain a stable nanosuspension, the selection of a stabilizer is of utmost importance, with potentially serious consequences for the system's physical stability such as Ostwald ripening, aggregation, sedimentation, and cake formation during milling and storage [5]. In this work, different concentrations of SDS were explored to achieve electrostatic stabilization.

The average particle size and polydispersity index of fresh corn starch nanosuspensions after storage for one month are given in Table 1. Moreover, the rheological properties are shown. The initial particle size of starch suspensions before milling operations was 15 μm. As the surfactant concentration increases, the stability of the particle size is significantly enhanced ($p < 0.05$), as suggested by the lower values of Z_{av} in fresh nanosuspensions and after storage. The PDI values are also significantly improved with the highest addition of surfactant. The increased electrical charge of nanoparticles, given by the ζ-potential values, might account in part for the enhanced stability [5]. As also shown in Table 1 and Figure 1, the increased concentration of SDS also enhanced the viscosity ($p < 0.05$) of nanosuspensions, providing an additional stabilization mechanism [7].

Table 1. Main properties of corn starch nanosuspensions.

Nanosuspension	K	n	ζ	Z_i	Z_f	PDI_i	PDI_f
7% *w/v* milled starch	0.019 a	0.81 a	−9.06 a	457.8 a	1356.5 a	0.34 a	0.65 a
7% *w/v* milled starch + 0.07% *w/v* SDS	0.008 a	0.77 a	−17.20 b	204.0 b	582.0 b	0.26 b	0.58 a
7% *w/v* milled starch + 1% *w/v* SDS	0.269 b	0.63 b	−27.35 c	344.6 c	457.8 c	0.43 c	0.34 b

K, consistency index (Pa.s^n); n, flow index; ζ, zeta potential (mV); Z, mean particle size (nm); PDI, polydispersity index; and "i" and "f" in lowercase letters stand for initial and final, respectively.

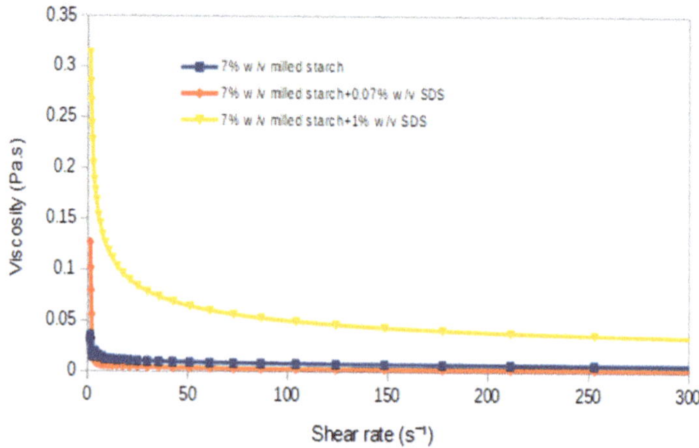

Figure 1. The viscosity of the corn starch nanosuspensions used to develop Pickering emulsions.

Regarding the interaction between starch and SDS, it is presumed that amylose forms left-handed helices with an internal hydrophobic cavity where ligands can reside. Although amylose has the ability to form complexes with diverse compounds, this process seems to be more favorable for charged salts of different fatty acids. These ligands produce an amylose non-retrogradable complex that remains soluble in water due to ionic group repulsion [11]. The SDS molecule exhibits a steric and functional similarity to these salts of fatty acids; therefore, this mechanism could be involved in the stabilization of nanosuspensions and emulsions. Indeed, the most stable oil droplet size in fresh emulsions was detected for the highest concentration of SDS, as can be seen in Figure 2. Immediately after their preparation, significant coalescence was observed with 0 and 0.07% *w/v* SDS. Those emulsions with the highest stability exhibited mean oil droplet sizes of 280 ± 4 and 292 ± 5 nm after their preparation and storage (one month), respectively. The electrical charge value was −25.40 ± 0.40 mV.

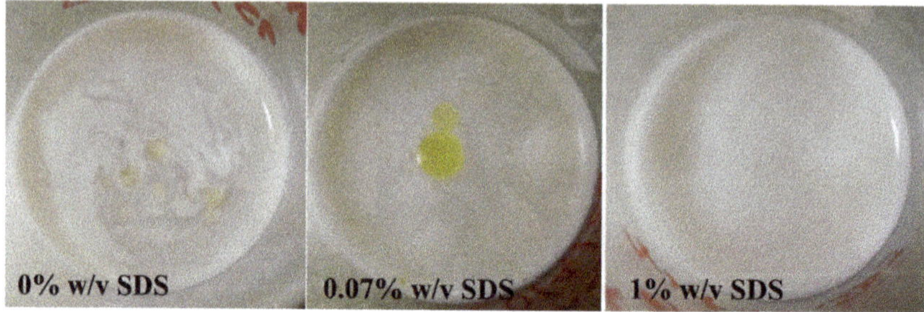

Figure 2. Photographs of Pickering emulsions containing chia oil and stabilized using corn starch nanoparticles obtained via media milling (7% *w/v* starch). From left to right: increasing concentration of SDS.

The formation of amylose complexes drastically modifies the properties of starch, improving its use in many relevant industrial applications. Amylose complexes have amphiphilic characteristics that reduce the interfacial tension at water/oil interfaces. These complexes also alter both the solubility and the emulsifying capacity with respect to amylose and the ligand. Moreover, when particles are adsorbed, they form a steric thick barrier on the droplet surface, which makes them resistant to destabilization phenomena such as coalescence and flocculation [11].

3.2. Characterization of Powders

Redispersible powders were prepared by spray drying nanosuspensions. The particle size and polydispersity index of powders immediately after spray drying and storage for one month are shown in Table 2. SEM micrographs of powders and the transmittance profiles after redispersion are given in Figures 3 and 4, respectively. Regarding the particle morphology, the structure of granules was significantly modified and the particles were swollen considerably as a result of milling. After spray drying, volumetric shrinkage was remarkably observed. In this context, the development of wrinkled morphologies in gelatinized starch particles after milling might obey the same mechanisms found for maltodextrins and whey proteins [5,8].

Table 2. Particle size of spray-dried powders.

Nanosuspension	Z_i	Z_f	PDI_i	PDI_f
7% w/v milled starch	2288.5 a	8058.5 a	1.00 a	1.00 a
7% w/v milled starch + 0.07% w/v SDS	373.9 b	8594.5 a	0.65 b	1.00 a
7% w/v milled starch + 1% w/v SDS	272.1 c	327.1 b	0.41 c	0.47 b

Z, mean particle size (nm); PDI, polydispersity index; and "i" and "f" in lowercase letters stand for initial and final, respectively.

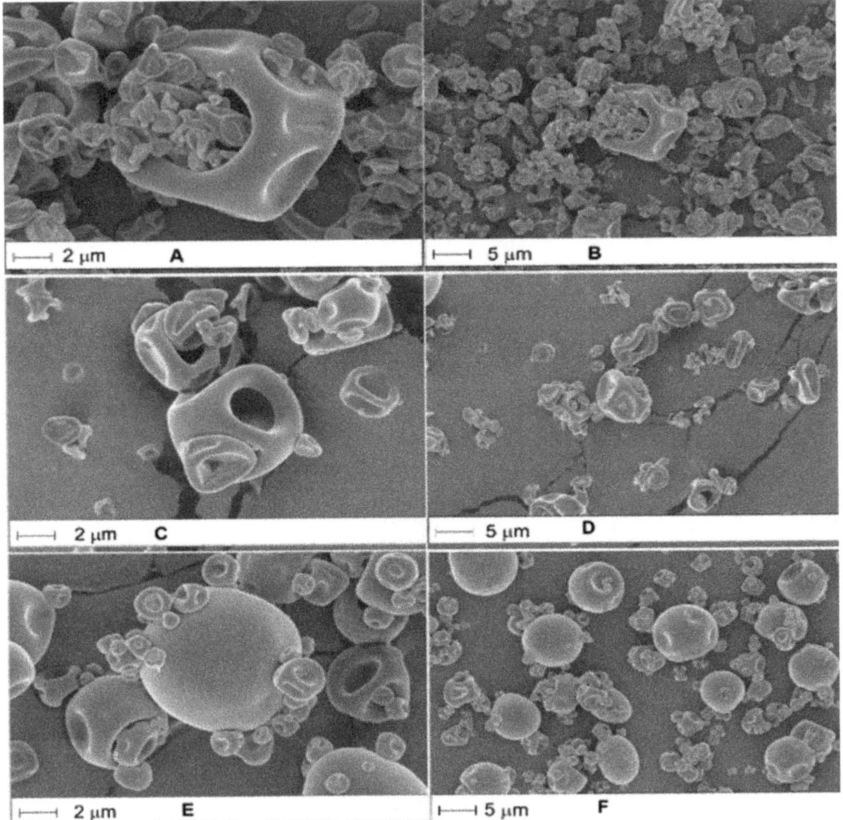

Figure 3. SEM micrographs of powders obtained from the spray drying of corn starch nanosuspensions; 7% w/v milled starch: (**A**) (5000×), (**B**) (2000×); 7% w/v milled starch + 0.07% w/v SDS: (**C**) (5000×), (**D**) (2000×); and 7% w/v milled starch + 1% w/v SDS: (**E**) (5000×), (**F**) (2000×).

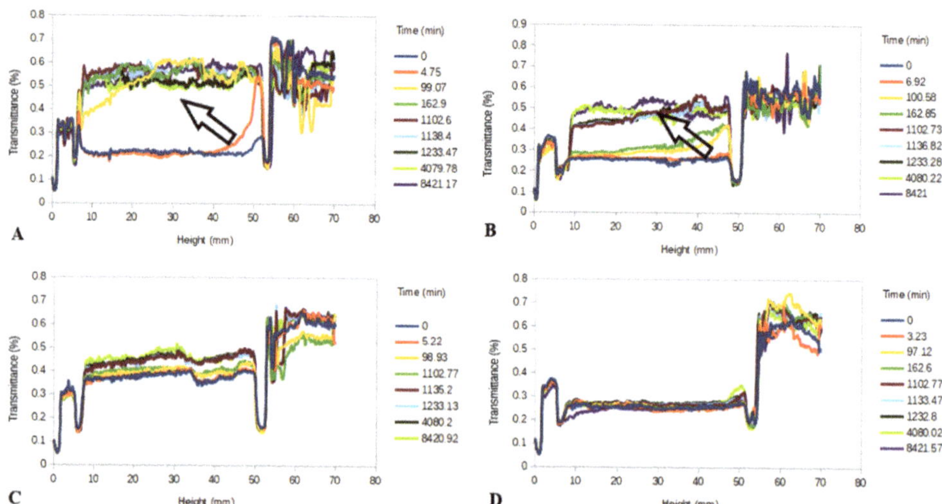

Figure 4. Transmittance (%) profiles of redispersed powders; (**A**) (native starch); (**B**) (7% *w/v* milled starch); (**C**) (7% *w/v* milled starch + 0.07% *w/v* SDS); and (**D**) (7% *w/v* milled starch + 1% *w/v* SDS).

According to Table 2, and with a similar tendency observed in nanosuspensions (Section 3.1), the highest concentration of SDS significantly enhances ($p < 0.05$) the dispersion of powders. In addition, the uniformity of the transmittance profile along the tubes during the measurement period is enhanced as the concentration of SDS increases. These profiles evidence a rapid sedimentation of particles for native starch and for milled starch without added surfactant, as indicated by the arrows in the figure (Figure 4A,B). Given the enhanced transmittance profiles of powders containing SDS, these were studied for their interfacial properties. Finally, as can be seen in Figure 5, a rapid decrease in the interfacial tension (down to 3 mN/m) was observed in those nanoparticles formulated with the highest concentration of SDS. The mechanisms explained in the previous section might account for these latter observations.

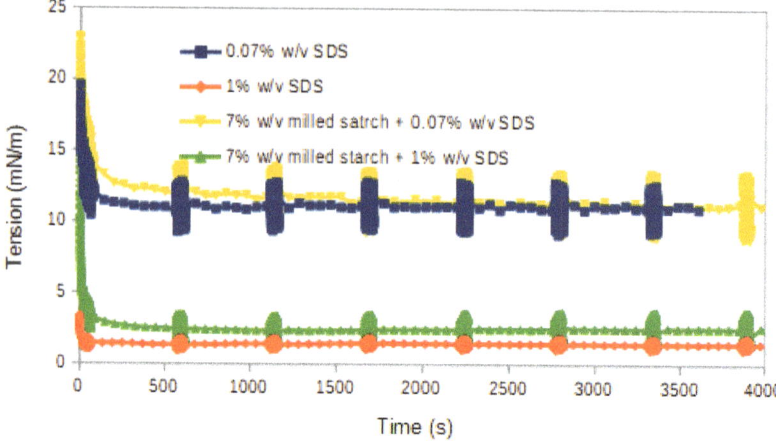

Figure 5. Interfacial tension determined at the chia oil/water interface using pendant droplet measurements.

4. Conclusions

The physical stability of corn starch nanosuspensions was significantly influenced by the concentration of surfactant (SDS). Moreover, the stability of oil droplet size during storage for one month in those Pickering emulsions containing chia oil in high loads might be explained by the adsorption and arrangement of starch nanoparticles at oil/water interfaces.

The results also demonstrated a promising stabilization of spray-dried Pickering emulsions for extending the shelf-life of chia oil within these delivery systems. Finally, this work shows that media milling can be utilized as a green methodology to develop sustainable nanoparticles able to deliver diverse bioactive compounds from different sources.

Author Contributions: Conceptualization, M.G.B., M.L.M., P.D.R., S.D.P., M.C.P., C.C.-S. and V.P.R.H.; methodology, M.G.B., L.L.-V., N.C., M.L.M., P.D.R., C.C.-S. and V.P.R.H.; software, M.G.B. and M.L.M.; validation, M.G.B., L.L.-V., N.C., M.L.M., M.C.P., C.C.-S. and V.P.R.H.; formal analysis, M.G.B., M.L.M., P.D.R., C.C.-S. and V.P.R.H.; investigation, M.G.B., L.L.-V., N.C., M.L.M., P.D.R., C.C.-S. and V.P.R.H.; resources, M.L.M., P.D.R., S.D.P., M.C.P., C.C.-S. and V.P.R.H.; data curation, M.G.B., M.L.M., P.D.R., S.D.P., M.C.P., C.C.-S. and V.P.R.H.; writing—original draft preparation, M.G.B., M.L.M. and P.D.R.; writing—review and editing, M.G.B. and M.L.M.; visualization, M.G.B. and M.L.M.; project administration, P.D.R. and S.D.P.; Funding acquisition, M.L.M., P.D.R., S.D.P., M.C.P., C.C.-S. and V.P.R.H. All authors have read and agreed to the published version of the manuscript.

Funding: This work was supported by the grant Ia ValSe-Food-CYTED (119RT0567), MinCyT-PIOBio (Decreto N° 349/2022-Res. Mi. N° 00000043/2022), CONICET (PIP 2021–2023 11220200102830CO), ANPCyT (PICT 2019-01122), and SeCyT-UNC, Cordoba, Argentina.

Institutional Review Board Statement: Not applicable.

Informed Consent Statement: Not applicable.

Data Availability Statement: The data presented in this study are available upon request from the corresponding author.

Acknowledgments: The authors would like to acknowledge the support of Ia ValSe-Food-CYTED, MinCyT, CONICET, ANPCyT, and SeCyT-UNC, Cordoba, Argentina.

Conflicts of Interest: The authors declare no conflict of interest.

References

1. Marcillo-Parra, V.; Tupuna-Yerovi, D.S.; González, S.; Ruales, J. Encapsulation of bioactive compounds from fruit and vegetable by-products for food application—A review. *Trends Food Sci. Technol.* **2021**, *116*, 11–23. [CrossRef]
2. Zhu, F. Starch based Pickering emulsions: Fabrication, properties, and applications. *Trends Food Sci. Technol.* **2019**, *85*, 129–137. [CrossRef]
3. Gonzalez-Ortiz, D.; Pochat-Bohatier, C.; Cambedouzou, J.; Bechelany, M.; Miele, P. Current trends in Pickering emulsions: Particle morphology and applications. *Engineering* **2020**, *6*, 468–482. [CrossRef]
4. Breitung-Faes, S.; Kwade, A. Mill, material, and process parameters-A mechanistic model for the set-up of wet-stirred media milling processes. *Adv. Powder Technol.* **2019**, *30*, 1425–1433. [CrossRef]
5. Bordón, M.G.; Martínez, M.L.; Chiarini, F.; Bruschini, R.; Camacho, N.; Severini, H.; Palma, S.D.; Ribotta, P.D. Preparation of corn starch nanoparticles by wet-stirred media milling for chia oil vehiculization. *Biol. Life Sci. Forum.* **2022**, *17*, 1. [CrossRef]
6. Martínez, M.L.; Marín, M.A.; Faller, C.M.S.; Revol, J.; Penci, M.C.; Ribotta, P.D. Chia (*Salvia hispanica* L.) oil extraction: Study of processing parameters. *LWT Food Sci. Technol.* **2012**, *47*, 78–82. [CrossRef]
7. Bordón, M.G.; Paredes, A.J.; Camacho, N.M.; Penci, M.C.; González, A.; Palma, S.D.; Ribotta, P.D.; Martínez, M.L. Formulation, spray-drying and physicochemical characterization of functional powders loaded with chia seed oil and prepared by complex coacervation. *Powder Technol.* **2021**, *391*, 479–493. [CrossRef]
8. Fu, Z.-Q.; Wang, L.-J.; Li, D.; Adhikari, B. Effects of partial gelatinization on structure and thermal properties of corn starch after spray drying. *Carbohydr. Polym.* **2012**, *88*, 1319–1325. [CrossRef]
9. Bordón, M.G.; Alasino, N.P.X.; Vilanueva-Lazo, A.; Carrera-Sánchez, C.; Pedroche-Jiménez, J.; Millán-Linares, M.C.; Ribotta, P.D.; Martínez, M.L. Scale-up and optimization of the spray drying conditions for the development of functional microparticles based on chia oil. *Food Bioprod. Process.* **2021**, *130*, 48–67. [CrossRef]

10. Felix, M.; Romero, A.; Carrera-Sánchez, C.; Guerrero, A. Assessment of interfacial viscoelastic properties of Faba bean (*Vicia faba*) protein-adsorbed O/W layers as a function of pH. *Food Hydrocoll.* **2019**, *90*, 353–359. [CrossRef]
11. Hay, W.; Fanta, G.; Felker, F.; Peterson, S.; Skory, C.; Hojilla-Evangelista, M.; Biresaw, G.; Selling, G. Emulsification properties of amylose-fatty sodium salt inclusion complexes. *Food Hydrocoll.* **2019**, *90*, 490–499. [CrossRef]

MDPI
St. Alban-Anlage 66
4052 Basel
Switzerland
www.mdpi.com

Biology and Life Sciences Forum Editorial Office
E-mail: blsf@mdpi.com
www.mdpi.com/journal/blsf